Encounters in Microbiology

Volume 2

Collected from *Discover* Magazine's "Vital Signs"

Edited by
Jeffrey C. Pommerville, PhD
Glendale Community College

JONES AND BARTLETT PUBLISHERS
Sudbury, Massachusetts
BOSTON   TORONTO   LONDON   SINGAPORE

World Headquarters
Jones and Bartlett
Publishers
40 Tall Pine Drive
Sudbury, MA 01776
978-443-5000
info@jbpub.com
www.jbpub.com

Jones and Bartlett
Publishers Canada
6339 Ormindale Way
Mississauga, Ontario
L5V 1J2
Canada

Jones and Bartlett
Publishers International
Barb House, Barb Mews
London W6 7PA
United Kingdom

Jones and Bartlett's books and products are available through most bookstores and online booksellers. To contact Jones and Bartlett Publishers directly, call 800-832-0034, fax 978-443-8000, or visit our website www.jbpub.com.

Substantial discounts on bulk quantities of Jones and Bartlett's publications are available to corporations, professional associations, and other qualified organizations. For details and specific discount information, contact the special sales department at Jones and Bartlett via the above contact information or send an email to specialsales@jbpub.com.

Copyright © 2009 by Jones and Bartlett Publishers, LLC

All rights reserved. No part of the material protected by this copyright may be reproduced or utilized in any form, electronic or mechanical, including photocopying, recording, or by any information storage and retrieval system, without written permission from the copyright owner.

Production Credits
Chief Executive Officer: Clayton Jones
Chief Operating Officer: Don W. Jones, Jr.
President, Higher Education and Professional Publishing: Robert W. Holland, Jr.
V.P., Sales and Marketing: William J. Kane
V.P., Design and Production: Anne Spencer
V.P., Manufacturing and Inventory Control: Therese Connell
Executive Editor, Science: Cathleen Sether
Acquisitions Editor: Shoshanna Grossman
Managing Editor, Science: Dean W. DeChambeau
Associate Editor: Molly Steinbach
Editorial Assistant: Briana Gardell
Senior Production Editor: Louis C. Bruno, Jr.
Senior Marketing Manager: Andrea DeFronzo
Cover Design: Kate Ternullo
Cover Images: Organism photo courtesy of P.J. Guard-Peter, colorization by Stephen Ausmus/USDA ARS; puzzle photo © Stillfx/ShutterStock, Inc.
Printing and Binding: Malloy Inc.
Cover Printing: John Pow Company

Library of Congress Cataloging-in-Publication Data
Encounters in microbiology : collected from Discover magazine's Vital Signs / [edited by] Jeffrey Pommerville and I. Edward Alcamo. — 2nd ed.
   p. cm.
   Includes index.
   ISBN 978-0-7637-5798-4 (vol. 1) (alk. paper) — ISBN 978-0-7637-5799-1 (vol. 2) (alk. paper)
   1. Medical microbiology—Case studies. I. Pommerville, Jeffrey C. II. Alcamo, I. Edward.
   QR46.E54 2008
   616.9'041—dc22   2008003027
6048
Printed in the United States of America
12 11 10 09 08   10 9 8 7 6 5 4 3 2 1

Contents

Preface...v
Introduction to the First Edition (Volume 1)...vii
The Steps Used When Diagnosing and
  Treating a Patient...ix

A Woman's Terrible Stomach Pain Turns Deadly...1
Tony Dajer

Bad Fever...6
Claire Panosian Dunavan

Is That Lump Malignant?...11
Mark Cohen

Microbes That Maim...15
Sheri Fink

Mystery Rash...21
Claire Panosian Dunavan

Why Can't He Walk?...26
Paul Austin

Bull's-Eye...31
Claire Panosian Dunavan

Can She Survive the Cure?...36
Stewart Massad

Gut Attack!...40
Tony Dajer

A Killer Raves On...46
Claire Panosian Dunavan

Why Are His Eyes Crossed?...51
Mark Cohen

The Sleeping Giant...56
Tony Dajer

Who's That?...61
John R. Pettinato

Just an Upset Stomach?...65
Claire Panosian Dunavan

Bad Blood...70
Mark Cohen

A Task in the Yard Turns Lethal...75
Claire Panosian Dunavan

Glossary...80

Index...90

Preface

Everybody loves a good mystery! There is something about a "whodunit" that draws us in, and we are not satisfied until we have identified the villain or solved the mystery. We love to be in the shoes of the professional or amateur detective trying to solve the crime or homicide using the clues collected during the investigation.

In *Encounters in Microbiology* we follow "real life" clinicians (general practitioners, specialists, or infectious disease physicians) who find themselves in the role of medical detective (à la TV's Gregory House) investigating a "whatdunit"; that is, the villain is a mysterious infection or other associated medical malady. Infectious diseases are still with us; in fact, they account for about 25 percent of disability-adjusted life years and some 25 percent of deaths worldwide. Even in developed nations, such as the United States, each year about 3 percent of outpatients are diagnosed with an infectious disease and almost 10 percent of all prescriptions are for antimicrobial drugs.

To accompany and complement the first volume of *Encounters in Microbiology, Second Edition*, we offer this second volume containing additional true stories drawn from the "Vital Signs" articles in *Discover* magazine. *Encounters in Microbiology, Volume 2*, presents true stories of relevant medical mysteries and patient diagnoses carried out by clinicians. When an ill patient presents to the clinician's office or clinic, the caregiver uses a set of steps to diagnose and treat the patient. These are described in The Steps Used When Diagnosing and Treating a Patient. Read through them before beginning the stories, as this initial reading will provide you with the medical detective techniques needed to make the stories easier to understand and follow.

As with the first volume, each encounter contains a set of Questions to Consider at the end of the story. Many of the questions stress critical reasoning skills. As such, they complement the epidemiological Textbook Cases found in my textbook *Alcamo's Fundamentals of Microbiology* (Jones and Bartlett Publishers). The answers to the questions are available through your Jones and Bartlett sales representative.

Encounters in Microbiology, Volume 2, also contains a Glossary and an Index. There are many specific medical terms (jargon) used by clinicians in their diagnosis of a patient. These terms usually identify a condition or malady affecting the patient; therefore, to help you decipher and understand these terms, a glossary has been added at the back of the book. If a term is not spelled out in the story—or even if it is—it almost certainly can be found in the glossary.

I wish to thank Cathleen Sether, Executive Editor, Science, at Jones and Bartlett Publishers for getting the ball rolling on this volume and Dean DeChambeau, Managing Editor, Science, also at Jones and Bartlett, for his continuing guidance and management of the project. I also want to thank Briana Gardell for her assistance in the initial preparation of the manuscript for this second volume.

The original *Encounters in Microbiology* was edited by the late Ed Alcamo. I include next his Introduction to the first volume. I hope you "enjoy your encounter in the world of microorganisms," and find the new additions to the medical mysteries useful and enjoyable.

Enjoy the medical sleuthing!

Jeffrey Pommerville, Ph.D.

Introduction

Microbiology can be a demanding science, with a host of new terms, a bevy of new theories, and a plethora of new insights. But it warms the soul and excites the spirit when we see practical applications for all we've learned. Placing microbiology into a useful and contemporary perspective seems to make the learning worthwhile.

And that's what *Encounters in Microbiology* is all about. In this volume, we present a series of stories involving real people and their encounters with microorganisms. In each case, the individual has reported to a hospital because he or she is suffering from a serious illness. Now it is up to the hero of the story, the physician, to find out what is going on in the patient and what is available to help. For many patients, the encounter with microorganisms ends happily, but sometimes, unfortunately, the ending is sad.

We now invite you to enter the real world of microbiology and experience real life, real people, and real cases. This volume was conceived by Dean DeChambeau, Managing Editor at Jones and Bartlett Publishers. All the articles in this volume were originally published in *Discover* Magazine, and all are true, although some of the names have been changed to protect the privacy of the patients. The articles have been written by some very talented physician-writers, and we thank them for their permission to use their stories. We also thank the editors of *Discover* Magazine for permitting us to reprint the articles.

And we thank you for giving us the opportunity to tell these stories. We hope you will enjoy your encounter in the world of microorganisms and will come to appreciate the people who fight them.

Best wishes,

E. Alcamo

The Steps Used When Diagnosing and Treating a Patient

The identification of the nature and cause of a patient's illness or disorder by a clinician is called a **diagnosis**. When the ill individual comes to the clinician's office or medical clinic, a series of diagnostic steps (**Figure A**) are set in motion. This includes a patient interview and an evaluation of the patient's reported symptoms, the physical examination findings determined by the clinician, the results of various laboratory and medical tests, and any other procedures pertinent to the investigation. If the patient's illness appears to be caused by an infectious agent, then the investigation may need to identify the causative agent and characterize the severity of the infection. When all this information has been evaluated and the clinician has reached a diagnosis, she or he can offer a **prognosis**, a prediction of the likely outcome of the disease. From this, a treatment procedure can be started and preventative (and possible public health) action initiated.

Because we are specifically interested in infectious diseases and disorders, the first step is to determine the **exposure history** of the patient.

1. Exposure History

The clinician will conduct an exposure history interview as part of the patient's overall personal and family history. Presentation to a clinician includes the patient's current illness and an oral report of his or her subjective symptoms. With regard to exposure, the clinician must cover the following topics when interviewing and examining the patient: can the patient determine the time of onset; can the patient pinpoint the time and place of exposure (e.g., home, work, recent domestic/international travel); has the patient had past infectious diseases, vaccinations, or immunological impairments; can the patient identify possible exposure sources (other humans, animals, foods, or environment).

Steps Used When Diagnosing and Treating a Patient

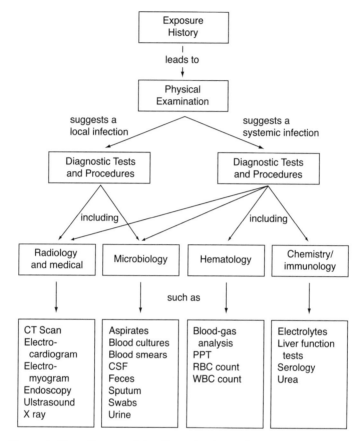

Figure A. Flow diagram of the diagnosis of an infection. CT = computed tomography; CSF = cerebrospinal fluid; WBC = white blood cell; RBC = red blood cell; PTT = partial thromboplastin time. (Modified from Greenwood, D., Slackl, R. C. B., and Peutherer, J. F. *Medical Microbiology*, 16th edition. London: Churchill-Livingston, 2002.)

The clinician's interview may indicate obvious signs or symptoms. For example, a child with a case of chickenpox (small, teardrop-shaped, fluid-filled vesicles on the torso) would be quite obvious to an examining pediatrician. In this case, little, if any, further investigation would be needed for a correct diagnosis, and the prevention of spread can be discussed. However, if this represents the beginning of an outbreak or epidemic, more specific public health measures may need to be initiated and a report made to state and national medical authorities. In other cases, signs and symptoms may not be so direct. The presence of a headache, fever, and malaise for two days could be symptoms for chickenpox as well as for a large variety of viral and bacterial infections.

In the interview with the patient, a review of body systems may uncover other parts of the body being affected by the disease. For example, coughing and shortness of breath may indicate respiratory system involvement. On the other hand, a burning on urination would suggest a urinary tract infection. All are part of the disease detective work done to locate the physical site of symptoms or what is called an **anatomical diagnosis**. Such a diagnosis may allow the clinician to narrow the list of possible infections or infectious agents.

2. The Physical Examination

As part of the **physical examination**, the clinician does a systematic examination of the patient, taking a blood pressure reading, measuring heart beat, and measuring body temperature. Specific emphasis is placed on the part of the body affected by the illness. For example, the throat, chest, and lungs are examined if a respiratory system infection is suspected.

The examination may allow the clinician to narrow the possibilities of diseases and/or infectious agents that would fit the clinical findings. The exposure history and physical exam may lead to the determination if the infection is local, such as the lungs or urinary tract, or systemic, involving several tissues/organs in the body. The clinician may then be ready to make a **differential diagnosis**, which narrows down the potential diseases to just those few that fit the clinical findings. If the presenting symptoms in a 45-year-old patient are a three-week cough, fever, chest pain, and coughing up blood or sputum, a differential diagnosis may include several respiratory infections but primarily tuberculosis (TB) as the cause. On the other hand, if the patient remembers a tick bite and has an expanding red rash at the bite site, the differential diagnosis is almost certainly Lyme disease, as few other arthropodborne diseases have these specific signs.

3. Diagnostic Tests and Procedures

As illustrated by many of the stories in this book, the first two stages—exposure history and physical examination—are carried out rather quickly, often on the initial interview.

However, the clinician may order one or more specific diagnostic tests to narrow down the short list of possible infections. Such tests or procedures can take some time and may be expensive. This is one reason why a clinician might attempt a final diagnosis through the physical examination or by using a minimal number of "standard" tests. For example, if TB is suspected, a chest x-ray or tuberculin skin test may be ordered to confirm the diagnosis. In addition, some diagnostic procedures may be noninvasive while others are invasive. Thus, comfort to the patient must be considered when diagnostic tests or procedures are being considered. Also, if someone such as a general practitioner is treating a patient, she or he may consult with an infectious disease specialist to obtain an expert opinion. Even Google searches have been used to help with patient diagnoses!

In some cases it may not be necessary to identify the specific pathogen as part of the diagnosis. For example, if all the diseases or agents identified by a differential diagnosis would be treated in the same way, or not at all (e.g., common cold, measles), there probably is no need to identify the pathogen. On the other hand, sometimes it is necessary to identify the actual causative agent of the infection. This **etiological diagnosis** may be important especially if it is a particularly dangerous disease. Again, taking TB as an example, drug-resistant TB is increasing rapidly worldwide. Therefore, it might be necessary to determine the drug resistance of the particular TB strain infecting the patient. In this case, a sputum sample would be taken for culture and growth of the bacteria. Then an antibiotic resistance determination would be made of the bacterial strain.

During the differential and etiological diagnoses, the clinician needs to be aware of a number of important epidemiological issues, especially if it is a disease like TB. The clinician needs to know if other individuals are at risk. Do particular behaviors (traveling, working in crowded places, etc.) expose one to the disease, and has the patient engaged in these behaviors? What is the geographical distribution of the disease, and has the patient been in these locales? Have there recently been additional cases reported locally? Has the patient been immunized, if possible, against this disease? During the patient interview and examination, many of these questions may be answered by the patient, assuming (which one often cannot) that the patient's ability to "self-report" is honest and accurate.

Although diagnoses and diagnostic tests obviously demand good judgment on the part of the clinician, for some conditions, written flow diagrams called **decision trees** (or algorithms) exist for making diagnostic decisions and for treating the patient; in other words, "If the patient has this, do the following test." Medical and health insurance companies often use diagnosis and treatment algorithms. An example of a decision tree for evaluating a suspected TB patient is illustrated in **Figure B**. As you can see, they can be quite extensive.

4. Treatment

Once the clinician has reviewed all the clinical information and diagnostic tests, hopefully a correct diagnosis can be made and the prognosis issued. Then treatment can begin. Note: Often as a precautionary measure, treatment may begin while diagnostic tests are being run.

There are two possible treatment scenarios. In **symptomatic treatment**, the clinician treats symptoms, such as pain, fever, cough, or muscle aches accompanying the underlying disease. Pain relievers, antihistamines, or cough suppressants may be prescribed for something like a cold or the flu. These treatments are simply supportive, making the patient feel better without influencing the final outcome or progression of the disease.

In a **specific treatment**, the clinician is specifically treating the diagnosed disease and hopefully affecting the final outcome. Typical specific treatments might be prescribing an antibiotic for a sinus infection or several antibiotics for something like TB. Antiviral agents, such as acyclovir, might be prescribed for shingles, although in actuality it is only treating the symptoms. Hopefully through proper treatment, the patient will progress through a period of disease decline and complete convalescence. However, less optimistic outcomes due to deadly pathogens sometimes occur, as some of the stories will describe.

xiv Steps Used When Diagnosing and Treating a Patient

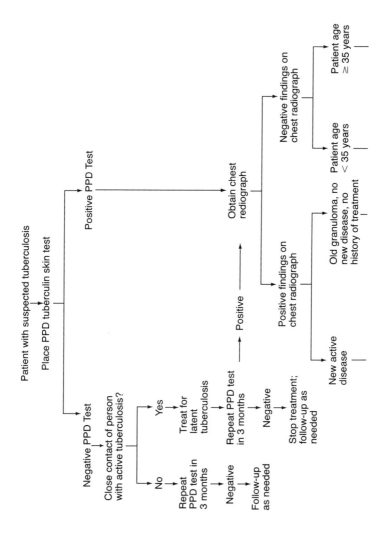

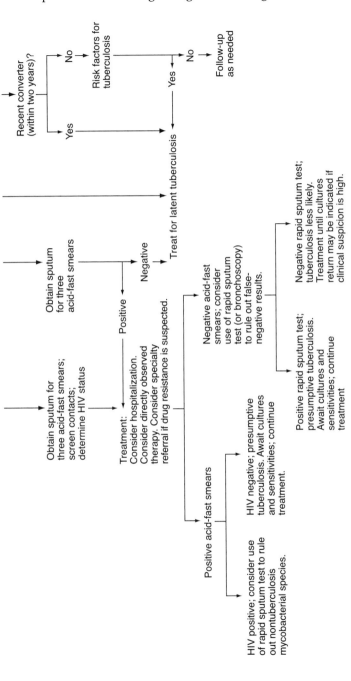

Figure B. Flow diagram for evaluation suspected tuberculosis patients. PPD = purified protein derivative; HIV = human immunodeficiency virus. (Modified from Jerant, A. F., Bannon, M., and Rittenhouse, S. Identification and management of tuberculosis. *American Family Physician* 61(9): 2667–2682, 2000.)

Antibiotic resistance is one of the major challenges facing the medical community today. Infections by an antibiotic-resistant bacterium can be more difficult to treat, and, as this encounter demonstrates, sometimes the simplest of infections can have unexpected and tragic results if treatment is delayed.

A WOMAN'S TERRIBLE STOMACH PAIN TURNS DEADLY

TONY DAJER

"How could she be so sick?" Mr. Kovacs implored, his eyes filling with tears. "We would have come in sooner."

His wife of 40 years lay on a stretcher, barely conscious, moaning in pain, a large blue diaper wrapped around her lower body to capture the now-incessant diarrhea. Her blood pressure hovered around 80, and her skin had turned sallow and pasty.

The surgeon pressed softly on the patient's abdomen. No matter where he probed, Mrs. Kovacs moaned louder. He stepped to the X-ray viewing screen. The abdominal CT scan showed a massively thickened large intestine.

"We need to get her to the operating room," he said. "She's obviously septic. Worse, the colon is probably dying."

Mrs. Kovacs had complained of diarrhea and stomach cramps for four days, but what finally brought her in was the weakness. "She can barely move," her husband had told the triage nurse. A healthy 65-year-old, Mrs. Kovacs had seen the inside of a hospital only to deliver her babies. As for doctors, she had needed their services for mild high blood pressure.

Reprinted with permission from the October 1999 issue of Discover *magazine. Copyright © Discover Magazine. All rights reserved. For more information about reprints from* Discover, *contact PARS International Corp. at 212-221-9595.*

2   A Woman's Terrible Stomach Pain Turns Deadly

The diarrhea had started out watery, not bloody, not too copious, with no fever. According to her husband, she didn't have any risk factors that might explain the persistent diarrhea. She hadn't been out of the country, she hadn't eaten any spoiled food, and she hadn't taken any antibiotics lately.

Two IVs dripped saline in, but her blood pressure would not rally.

"This morning she was walking around," one of her daughters said fearfully. "We thought she had a stomach flu."

The lab results had come back sky-high, with a white blood cell count of 25,000 (normal is 4,000 to 11,000). Her diarrhea had made her so dehydrated that her kidney function was one-third of normal.

The surgeon, still puzzled as to how a gastrointestinal infection could fell a healthy woman, went back over the history.

"No antibiotics in the past few months?" he asked. "You're sure?"

Another daughter had since arrived. "Oh, yes," she exclaimed. "She had a tooth infection about three weeks ago. The dentist gave her something. I brought the container."

She fished in her pocket and handed a plastic vial to the surgeon.

"Clindamycin," he read aloud.

Mr. Kovacs, understandably, had forgotten. His wife had finished the drug treatment two weeks earlier.

"This finally makes sense," the surgeon said. Then, as gently as he could, he addressed the family. "She's in for a very rough time."

In the 1960s, reports of a bizarre and sometimes lethal colon affliction appeared in the medical literature. Because the cell debris and inflammatory gunk that lined the colon looked like a yellow-green membrane, researchers called it pseudomembranous colitis, but its direct cause remained elusive. Clear from the start, however, was that antibiotics—clindamycin in particular—were implicated.

The human colon harbors a complex ecosystem of bacteria. By and large, our bacterial companions behave like a big happy family in which all mind their place and do their part. Some of the bacteria use oxygen; some don't. Many aid in digestion and make nutrients like vitamin K. The social order is fragile, however. Add antibiotics and the good bacteria die, allowing nasty competitors to move in. The most common side effect is diarrhea. Most cases occur because the bacteria-depleted intestine cannot fully digest carbohydrates, and the unabsorbed sugar provokes the runs.

Pseudomembranous colitis is different. In 1978 researchers traced the cause to toxins made by the anaerobic bacterium *Clostridium difficile*. The toxins not only trick intestinal cells into secreting massive amounts of fluid but can also fatally gum up their protein-making machinery. Symptoms range from none (a healthy carrier state) to simple diarrhea to toxic mega-colon, where the colon balloons in size to become an inflammatory, necrotic cesspool. The older and more debilitated you are, the more likely you'll suffer the most severe effects.

Pain and fever caused by *C. difficile* may begin from 2 to 10 days after antibiotics are started or up to 10 weeks after they're stopped. Clindamycin leads the pack in terms of risk, probably because it's so good at wiping out the dominant, friendly anaerobic, or oxygen-shunning, bacteria (many antibiotics attack only aerobic bacteria). But common antibiotics like cephalosporins (Keflex and Rocephin) and penicillins (amoxicillin and ampicillin) also cause their share of cases. Recently another widely used class of antibiotic, the fluoroquinolones (which include Cipro and Levaquin), have shown worrisome signs of catching up. Adding to the dilemma are some studies suggesting that stomach acid–reducing drugs—among the most widely prescribed medications in the world—may also increase the risk of *C. difficile* disease.

Up to 8 percent of healthy adults harbor *C. difficile* in their guts. Always the opportunist, the bug thrives in hospitals, where it infects 3 million patients a year—including about 13 percent of all inpatients who spend up to two weeks in the hospital. Mrs. Kovacs was an unlucky outlier—only 20,000 outpatient cases are reported each year in the United States. That's the good news. The bad news: The prescribing of antibiotics for everything from colds to sinusitis and benign coughs seems to have spawned another superbug. Over the past five years, a new strain of *C. difficile* that produces 20 times the usual amount of toxin has blitzkrieged through hospitals and nursing homes in Canada, the United States, and Europe, killing up to 10 percent of its elderly victims.

In hospitals, alas, *C. difficile* spreads mostly via health care workers. The new trend toward relying on regular squirts of alcohol-based gels to clean the hands might be making things worse because the gels do not kill the bacterium's spores. Old-fashioned handwashing and isolation do. Among Quebec hospitals hard-hit by a recent outbreak, enforced washing with good old soap and water dropped the infection rate by half.

On the outpatient front, Great Britain has seen the incidence of *C. difficile* infection skyrocket from less than one per 100,000 people in 1994 to 22 per 100,000 in 2004. The profile of the patient is changing too. In 2005 the Centers for Disease Control received reports of eight healthy outpatients in the United States who suffered serious *C. difficile* disease and hadn't taken antibiotics in the preceding three months. In other words, some strains of the organism are muscling aside good gut bacteria even without our help.

Ironically, the treatment for *C. difficile* disease is more antibiotics: The idea is to pare back the runaway intruder with anaerobic-specific antibiotics like metronidazole or vancomycin. Although they kill good anaerobic *Bacteroides fragilis* as well, they allow normal aerobes like *Escherichia coli* to regain a foothold and begin to restore ecological harmony.

For Mrs. Kovacs, the surgeons tried every intervention: high-dose antibiotics, fluids, and pressors—medications that boost blood pressure. Still, her vital signs continued deteriorating. The next morning, hoping to extinguish the source of bacterial toxins and the corrosive by-products of massive cell death, the surgical team removed her colon. But the defenses kept crumbling. Two days later they had to take out a portion of small intestine. Then came a grim procession of secondary complications: a gallbladder infection, pneumonia, and internal bleeding.

"I don't think she'll make it," the surgeon confessed to me about two weeks into her treatment.

He fought a long, hard rearguard action, aggressively working up and treating every new complication. I didn't ask if he was fighting so hard because this once hale, vigorous woman was at death's door due to a prescription for a tooth infection.

Two weeks later, with her family on a round-the-clock vigil, she succumbed. Had she come in sooner, her death might have been averted. Maybe a more adamant warning from her dentist about clindamycin's potential dangers would have saved her life. But it would be hard to pin blame only on him, given that we American doctors still uselessly prescribe antibiotics—to the tune of over 10 million prescriptions a year—to patients with viral upper respiratory infections.

And we still think we're doing good.

UPDATE

Since 2006, Great Britain has been dealing with more than 6,300 hospital-based "superbug" infections in many of its government-run National Health Service facilities. One of these superbugs is Clostridium difficile. *In October 2007, the Los Angeles Times reported that British health authorities suspected more than 90 people had died from C.* difficile *infections. The bacterial infection apparently spread through three hospitals in south-central England after patients were forced to defecate in their beds and then wait for hours for clean sheets. Since bacteria are found in feces, it is not surprising that patients can become infected if they touch items or surfaces that are contaminated with feces and then touch their mouth or mucous membranes. Throughout Europe, outbreaks of C.* difficile *hospital infections are increasing. In fact, an additional 255 deaths have been at least partially linked to C.* difficile. *The British government has pledged 280 million dollars to combat C.* difficile *transmission and infection by disinfecting and cleaning every British hospital.*

QUESTIONS TO CONSIDER

1. Why is the illness affecting Mrs. Kovacs called pseudomembranous colitis?

2. What has "spawned" *Clostridium difficile* as another superbug?

3. What was the reason for prescribing anaerobe-specific antibiotics to treat Mrs. Kovacs?

4. What was the source of Mrs. Kovacs's infection?

5. How is *Clostridium difficile* usually spread in hospitals, and how can the infection rate be reduced?

When we travel abroad, we often think of the great adventure that lies ahead. Often we do not consider what health risks may be inherent in that travel, especially if it is to a "health-safe" region of the world. Unfortunately, as this encounter relates, sometimes the great adventure can be more than we bargained for.

BAD FEVER

CLAIRE PANOSIAN DUNAVAN

The message was brief but urgent: "Honeymooner arriving tonight from Tahiti. Has fever, headache, hemorrhagic rash. Will take ambulance to our emergency room. Can you meet her there?"

That year, I took care of three returning newlyweds with the same tropical woe—but none as sick as Susie Gold.

Susie and her fiancé, Jeff, had resolved to have a one-of-a-kind wedding in the South Pacific, where they would exchange vows under a cabana of palm fronds. Everything went off without a hitch, but five days later Susie developed teeth-chattering chills. Her muscles felt sore and bruised. At first she downplayed her symptoms, figuring she was wiped out from the wedding or fighting a bug she had picked up on the plane. It would pass.

Sure enough, a day later, she was better. She and Jeff resumed snorkeling and strolling along the beach, their happiness marred only by sunburned noses and mosquito bites.

The next morning Susie's head and eyeballs ached furiously, her brow was hot, and her leg sported a rash that looked like purple shooting stars. Her biggest scare came after flossing her teeth. She tasted salt, looked in the mirror, and saw bright red blood.

Reprinted with permission from the June 2002 issue of Discover *magazine. Copyright © Discover Magazine. All rights reserved. For more information about reprints from* Discover, *contact PARS International Corp. at 212-221-9595.*

Jeff called Susie's dad, a psychiatrist. He listened to Susie's symptoms, then rang off to consult with the family internist. Thirty minutes later he was back on the line.

"Jeff, you've got to bring Susie back. Don't worry; I've arranged everything. Once you land in Los Angeles, an ambulance will take her straight to the university hospital."

If Dr. Gold had actually spoken to me before Susie boarded her plane, I probably would have vetoed the daylong wait before she was seen by a doctor. To an infectious-diseases specialist, her constellation of fever, headache, pinpoint bleeding in the skin indicates meningococcemia until proven otherwise. Meningococci are bacteria that invade the blood and the meninges that line the brain. A 12-hour delay in receiving antibiotics can be the difference between life and death.

But deep in my gut, I worried that Susie might have another misery that can mimic meningococcemia. Starting in the 1980s, dengue—a mosquito-borne virus that dogged Allied and Japanese troops during World War II—made a stunning comeback in the Pacific, Southeast Asia, and the Caribbean. Like meningococcal infection, dengue also causes fever, headache, and hemorrhagic rash.

A small, black-and-white mosquito, *Aedes aegypti*, is dengue's main vector. It loves to breed in water-filled crockery, cisterns, trash containers, and spare tires that surround human habitation in the tropics. That's a lucky coincidence for an insect whose survival depends upon blood meals. In turn, the blood-borne dengue virus profits from its vector's feeding habits. With the slightest provocation, the female *Aedes* stops her blood meal, only moments later to resume probing and siphoning the same or another nearby victim. Thus, a lone mosquito carrying dengue from a previous host can spread the virus to multiple recipients.

After an *Aedes* loaded with dengue inoculates its victim, illness begins within seven days. Although symptoms range widely, fever, chills, and headache often herald the attack, along with facial flushing, swollen glands, and a mild sore throat. Then comes a deceptive lull before dengue's encore fever arrives. This later stage features an array of skin rashes as well as dengue's famous aches, otherwise known as breakbone fever.

Dengue's most dreaded complications are hemorrhage and shock, which typically strike children and adolescents battling the infection for a second time. During a second bout of dengue, old antibodies are thought to bind to the new virus, but they fail to clear it because of changes in the new infecting virus. Instead, antibody-virus complexes are engulfed by watchdog cells called macrophages. During severe infections, macrophages can release chemical signals that cause capillaries to leak, which results in bleeding and, at times, drastic depletion of plasma volume.

Still, Susie's case didn't stack up perfectly for dengue because her bleeding didn't fit with a first-time infection. A true diagnosis would be tricky: There's no easy test for dengue.

Eight hours later I got a call on my pager from our emergency room. Susie had arrived.

I hurried to the cubicle where she lay pale and sweaty but lucid. Her blood pressure was low and her pulse was high. She had bleeding gums and a sprinkling of hemorrhagic dots on all four extremities, just as she had described. Otherwise, the physical exam revealed only a few lentil-size lymph nodes and a tender liver edge below the right rib cage.

"She didn't eat a thing on the plane," Jeff said, his voice tinged with worry, "and when she stood up at the end of the flight, she nearly fainted."

Drip, drip, drip: An intravenous set was running saline into Susie's vein at the fastest rate the tubing would allow. Adding fluid would help bring up her blood pressure and prevent her from going into shock.

"Her vascular volume is low," I replied, avoiding the term "shocky," which the ER resident had used to describe the washed-out bride. "Right now she needs lots of fluid. We'll deal with food later."

I turned to Susie. "Don't worry. It will be rough, but you'll pull through."

At that point I left to review lab tests. Just as expected in a major viral assault, Susie's leukocytes and platelets were low. That meant she was depleted of cells that counter infection and bleeding. At the same time, her liver enzymes were elevated three- to fourfold, corresponding to the swollen, tender edge I felt on exam. Everything fit with dengue, yet nothing was truly diagnostic. Meanwhile, it would take 48 hours before blood cultures were known to be negative for meningococcus.

Sometimes being a purist is a mistake in medicine. Although I was confident that Susie had dengue, a disease for which antibiotics are ineffective, I asked myself: In her shoes, would I want antibiotic treatment until a serious bacterial infection was 100 percent excluded? My answer was yes.

"Let's start ampicillin and ceftriaxone for now," I said to the resident, "but don't forget to order an antibody test for dengue fever. It may take a long time to get the result, but it could be our only proof."

My next stop was the library. All day a vague memory had nagged me. Hadn't I seen a case report describing hemorrhage and shock in first-time dengue? Or was my mind playing tricks on me?

The American Journal of Tropical Medicine and Hygiene came through: "Dengue shock syndrome in an American traveler with primary dengue 3 infection" (March 1987). In the 15 years since that report, with the blossoming of dengue and exotic overseas travel, such cases in tourists are no longer rare. For reasons that are unclear, even these first-time infections can occasionally lead to bleeding and shock.

As for Susie, she spent the next few days in the hospital while her plasma volume restored, her bleeding stopped, her skin hemorrhages cleared, and her lab tests normalized. As expected, her blood cultures remained negative, so we finally shut off her antibiotics 48 hours after admission. Three days later, when she and Jeff were ready to return to the East Coast, the only remnant of her ordeal was depression. Post-dengue blahs, sometimes referred to as neurasthenia, were well known to British colonials. I reassured Susie that her mood was normal and would gradually improve on its own. The following week her dengue antibody sent from the emergency room finally came back positive.

UPDATE

Today, dengue fever (DF) is the most important mosquito-borne viral disease affecting humans. In fact, its global distribution is comparable to that of protozoan-caused malaria and the World Health Organization (WHO) estimates that some 2.5 billion people live in areas at risk for epidemic transmission. Each year, millions of cases of DF and thousands of cases of dengue hemorrhagic fever (DHF), the more severe form of the disease, are reported by the WHO. As this

10   Bad Fever

encounter illustrates, most dengue cases in American citizens occur as a result of travel to a dengue-endemic region of the world. However, dengue is now reaching epidemic levels in the Caribbean and in Latin America. The Pan American Health Organization (PAHO) says that changing climate and cyclical weather patterns, increased tourism, and migration have stimulated increased prevalence. The PAHO expects more than 1 million cases in the Western hemisphere in 2007 with close to 20,000 cases of dengue hemorrhagic fever, which will kill several hundred.

QUESTIONS TO CONSIDER

1. What was it about Susie's infection that didn't fit with the typical symptoms of the disease?
2. If antibiotics are useless in treating Susie's disease, why were they initially used?
3. Meningococcemia was initially considered as the diagnosis. What signs and symptoms do meningococcemia and dengue fever share?
4. How did Susie become infected with the viral pathogen?
5. Suggest a reason why dengue is sometimes called breakbone fever.

Children are always getting cuts and scratches. Such injuries are usually very minor and not of concern to the health of the child, but sometimes a swollen lump can be of concern and lead a physician to consider more serious consequences. Therefore, getting a correct and truthful patient history can make a diagnosis easier. As this encounter shows, getting such a truthful history can be difficult—especially if coming from a child.

IS THAT LUMP MALIGNANT?

MARK COHEN

My second-year resident ducked his head into my office in the pediatric clinic. "Hey, Dr. Cohen," he said, "can you come and look at a girl with a lump on her arm?"

The area on the 7-year-old girl's upper right arm had been swelling for about a week, he said, and it was getting bigger. She'd had a slight fever about a week ago, but none since. The lump was painful and hard, and yesterday the mother had noticed some slight redness of the skin. The child had been healthy, and her immunizations were up to date. She wasn't taking any medications, and no one at home had been ill. The resident noted that she hadn't had any injury, such as a bite, or any exposure to animals.

We went in to see the girl. I introduced myself and crouched down beside her.

"Hi. Can I check your arm?"

She held out her arm, a wary look on her face. I began by examining her wrist and forearm, carefully feeling the tissue.

Reprinted with permission from the June 2005 issue of Discover *magazine. Copyright © Discover Magazine. All rights reserved. For more information about reprints from* Discover, *contact PARS International Corp. at 212-221-9595.*

12 Is That Lump Malignant?

Kids' lumps and bumps are usually benign. But sometimes they prove to be diagnostic challenges, and every once in a while they turn out to be a sign of something very serious. When a child comes in with a lump, the remote but real possibility of a tumor is always in the back of my mind. A malignant tumor often feels rock hard, as opposed to an enlarged lymph node, which is usually firm but not hard, or a cyst, which might feel soft.

"Does that hurt?" I asked. She shook her head. "Good. Can you show me where the bump is?" She pointed to the inside of her upper arm. I could see the swelling, right where the biceps muscle makes contact with the bone. My fingers slid lightly over the surface of the faintly reddened spot, then probed a little deeper. The child winced a bit, but she did not seem to be in great pain.

The swollen area was about the size of a matchbook. It felt too hard to be a benign enlarged lymph node, and it didn't move freely under pressure as a lymph node would. Enlarged nodes tend to crop up in the armpit or around the elbow, not in the middle of the upper arm. The lump's redness and tenderness suggested an infection, but its firmness suggested a tumor.

The resident and I decided to take blood tests and start the girl on antibiotics in case it was a local infection. We also ordered an ultrasound examination of her arm. He would see her back in the clinic in the afternoon.

The next day the radiologist informed us that the child's lump was a lymph node, and she suspected it was enlarged due to cat scratch disease. "Cat scratch disease?" I said. "That's a surprise. I'll be right there."

Once considered rare, cat scratch disease is now known to be one of the most common causes of swollen and infected lymph nodes in children. The causative bacterium, *Bartonella henselae*, is commonly found in cats, especially kittens. The bacteria enter the body when a child is scratched or bitten by an infected feline. Within 3 to 10 days there may be a small red bump at the site of the scratch. Over the next few days the bacteria multiply, and the child may develop a fever, headache, or other signs of mild illness. Meanwhile, the bacteria move into the intricate network of lymph vessels, a system of channels that carry fluid, pathogens, and debris out of the tissues. The lymph node becomes large and firm as bacteria multiply and attract passing immune cells that make antibodies to the pathogen. Fortunately, cat scratch disease usually subsides without treatment.

The resident reassured the child and her mother that she would get better. He sent her to the lab for a blood test to confirm the diagnosis, which would be back in a few days. After they left, we went to see the radiologist. She pointed out the distinctive appearance of an enlarged lymph node on the ultrasound image. I realized that the extreme firmness of the node was the result of the infection's intensity and rapid progression. Still, there was one other part of the story that didn't make sense.

"I thought she hadn't had any exposure to animals," I said to the resident.

"Well, that's what she told me. Turns out she was scratched by the neighbor's kitten a couple of weeks ago."

"Let me guess," I said. "She didn't tell her mother, or us, because Mom told her not to play with the neighbor's kitten."

"That's exactly right!" he said. "How did you know?"

The radiologist and I exchanged smiles, and I turned to the young resident with a grin. "You don't have kids, do you?"

UPDATE

Nearly half of all domestic cats, especially kittens, at some point in their lives are infected with Bartonella henselae. *Cats do not get ill from the infection, which is spread between cats by infected fleas. However, to date there is no evidence that an infected flea can transmit the bacterium directly to humans. The illness occurs more often in the fall and winter, and, in the United States, there are some 22,000 cases reported annually. If one does get a scratch or bite from a cat, wash the site well with soap and water. For people with weak immune systems (e.g., cancer and organ transplant patients and people with HIV/AIDS), cat scratch disease may cause more serious problems. In AIDS patients, infection can lead to bacillary angiomatosis, which is a severe inflammation of the brain, bone marrow, lymph nodes, lungs, spleen, and liver; the disease can be fatal in HIV-infected individuals. In fact, bacillary angiomatosis is so characteristic today of AIDS that the Centers for Disease Control and Prevention (CDC) list it as an AIDS-defining disease. It can be easily treated with antibiotics such as erythromycin and doxycycline.*

14 Is That Lump Malignant?

QUESTIONS TO CONSIDER

1. What did the ultrasound examination of the child's arm indicate?
2. Why was a swollen lymph node initially ruled out in the child's diagnosis?
3. Besides the presence of the right arm swelling, what other signs and symptoms might the child demonstrate?
4. How did the child become infected with the bacterium?
5. From this encounter, about how long is the incubation period for cat scratch disease?

When an illness strikes an individual, the value of having frank discussions with the physician cannot be overemphasized. Many infections, such as abdominal pain, cannot be easily diagnosed by a physical exam of the patient, so the physician must depend on the patient for a complete and honest history. As this encounter makes clear, for adolescents, frank conversations with their parents are also important for sustaining harmony in the family, averting undue worry when disease does occur, and maintaining a healthy, disease-free body.

MICROBES THAT MAIM

SHERI FINK

"Here we go again," Ron Smith said with a sigh. "This is the third time we've been to the emergency room in five weeks."

He nodded toward his 15-year-old daughter. "Lorna's got a stomachache. She's throwing up. The last time she came here, it was a problem with her appendix."

Lorna lay on the stretcher, looking away from her father. She slid me a blasé look.

"The pain you're having now," I asked, "when did it start?"

"Two days ago. I thought it was from something I ate at a dinner."

"Did your friends get sick?"

She shook her head. "No."

"How is the pain this time compared to the other times you were here?"

"Not as bad," she said.

Reprinted with permission from the February 2001 issue of Discover *magazine. Copyright © Discover Magazine. All rights reserved. For more information about reprints from* Discover, *contact PARS International Corp. at 212-221-9595.*

The nurse's chart said Lorna's vital signs were normal and she had no fever. I placed my hands on her abdomen and began to press.

"Tell me where it's sore."

No response. Her belly felt soft against my fingertips, not hard as it does when a patient guards against pain or when an infection irritates abdominal muscles.

A complete exam yielded only two subtle abnormalities. When I listened with my stethoscope, the clicks and gurgles of her bowel sounds were less frequent than normal. Tapping the right lower side of her belly with my fingertips produced a dull sound. The sign meant something solid beneath the tissue, anything from feces to a mass.

And I noticed something else. Lorna and her father did not speak or even look at each other, and I wondered if this could offer us a clue.

In the ER, woman plus abdominal pain equals pregnancy outside the uterus until proven otherwise. Known as an ectopic pregnancy, it occurs when a fetus implants in the wrong place, often causing massive internal bleeding. An emergency physician's biggest responsibility is to rule out just such urgent, dangerous conditions.

I looked at Lorna, who seemed young, innocent, and irked. Awkwardness nearly prevented me from asking my next questions. "Is there any chance you could be pregnant?"

"No," she said.

Abdominal pain is one of medicine's most formidable diagnostic challenges. Most abdominal disorders lack specific symptoms and signs, and even advanced imaging techniques cannot always portray a condition as common as appendicitis. Answers often do not come until a patient undergoes surgery.

I called up Lorna's medical records on the hospital computer. Just as her father said, Lorna had visited the hospital twice over the past month for abdominal pain due to a suspected ruptured appendix. The report also indicated that although the suspected rupture had created a pocket of infection, the illness appeared to be subsiding. So the surgeons made the rare choice not to operate; instead, they gave Lorna a strong mixture of antibiotics. They planned to remove the diseased appendix several weeks later.

But Lorna's father had failed to mention that fluid collected from a pelvic exam during Lorna's first hospitalization had tested positive for the parasite chlamydia. That meant she was probably sexually active, because the parasite is almost always transmitted via sexual intercourse.

Doctors had treated Lorna with the antibiotic azithromycin to kill the chlamydia, but gynecologists suspected that the infection, rather than a ruptured appendix, might have caused Lorna's abdominal pain and hospital admissions.

The attending physician and I returned to perform a pelvic examination. Lorna had no masses and no pain when I manipulated her uterus and felt for her ovaries. I asked if she wanted to tell us anything without her father present.

"No."

Outside the room, the attending physician asked me what I thought. Lorna no longer had abdominal pain, she did not have a fever, and she was already on antibiotics. The possibility of a recurrent bacterial infection was unlikely. Given her sexual activity, we needed to rule out an ectopic pregnancy— unlikely too. I began to think Lorna might have been right about the bad meal.

Had the lab tests come out negative and Lorna kept down some food, we might very well have sent her home. But the attending suggested I notify the surgeons who took care of Lorna last time.

"I can't get down to see her for another half hour, " said the surgical resident. "Can you get started on X rays?"

I ordered them and went to lunch. When I returned, I found the surgeon preparing to admit Lorna. He asked for a CT scan, a radiological study that provides slicelike internal images of the body by gathering X-ray transmission data from many different directions.

The X rays showed a blockage in Lorna's small bowel. The surgeon planned to run a tube from Lorna's nose into her stomach to decompress the built-up fluid and gas. Her bowel function, he said, should return to normal in a few days, and she could go home.

The next morning, I learned that Lorna had not gone up to the patient floor. The results of the CT scan had sent her right into the operating room.

Despite the lack of a palpable mass on her slender frame, the CT revealed a huge abscess in the right side of her pelvis.

Her immune system had corralled microbes into a circumscribed area, so she failed to show signs of infection such as fever, faster heart rate, and intense pain. But the infected fluid had pressed on her bowels, causing tissues to stick together and obstruct fecal movement.

An abscess usually develops next to a diseased organ. When the surgeons drained the fluid and freed the adhering tissues, they found Lorna's abscess had originated not from her appendix but from the fallopian tube near her right ovary. The antibiotic had failed to cure Lorna's infection because it was so advanced. The cause of Lorna's medical troubles was now clear: chlamydia.

Chlamydia is the most commonly reported sexually transmitted disease in the United States, with an estimated 4 million new cases each year. Certain subtypes of the parasite *Chlamydia trachomatis* infect the epithelial cells lining the reproductive tract. The organisms grow inside the cells, killing them, and the body's immune response leads to inflammation and further local damage.

Chlamydia causes such mild initial symptoms that doctors detect most infections late or incidentally on routine testing. Delays in diagnosis and treatment increase the chance that the pathogen will spread up the reproductive organs to infect the cells that line the tubes and ovaries, causing pelvic inflammatory disease. About 20 percent of such patients end up with chronic pelvic pain. A similar number develop scarring in the fallopian tubes, which prevents conception. Lorna faced some tough possibilities.

I stopped in to see Lorna later that day. She talked a little, but clammed up when her parents appeared. When I left the room, the Smiths followed me.

"What caused this?" asked Mrs. Smith.

Despite Lorna's three hospital admissions, her parents had barely an inkling of their daughter's sexually transmitted infection. Doctors had respected Lorna's legal right to keep medical information about her sexual activity private. I urged her mother to speak with Lorna.

"I try, but I can't. When it was time to give her the menstruation talk a few years ago, she said 'I know it all' and ran out of the room. I was so relieved!"

Parents and physicians often fail to talk with teens about sexual activity and sexual disease transmission. Condoms reduce the risk of infection. Douching after sex increases it. And vaccinations against hepatitis B, which is sexually transmitted, are available.

"It's so hard to know what they're doing," Lorna's father said. "You can't keep them locked up, and we have to work." He looked down the hall. "Anyhow, I hope this will take care of it."

UPDATE

Chlamydia is the most frequently reported bacterial sexually transmitted disease in the United States. In 2005, more than 976,000 chlamydial infections were reported to the Centers for Disease Control and Prevention (CDC) from 50 states and the District of Columbia. Chlamydia is known as a "silent" disease because about 75 percent of infected women and about 50 percent of infected men have no symptoms. Thus, under-reporting is common since most people infected with chlamydia will not seek testing. The CDC estimates that almost 3 million Americans are infected with chlamydia each year. Women are frequently re-infected if their sexual partners are not treated. As in this encounter, an infection of a woman's reproductive system can lead to pelvic inflammatory disease (PID), which can cause infertility or serious problems during pregnancy. PID occurs in up to 40 percent of women with untreated chlamydia, and babies born to infected mothers can get eye infections and pneumonia from chlamydia. If one is sexually active, the risk of contracting chlamydia can be reduced by using condoms. The CDC and health experts recommend that women younger than 25 get a chlamydia test every year, one reason being that women infected with chlamydia are up to five times more likely to become infected with HIV, if exposed.

QUESTIONS TO CONSIDER

1. What was the original diagnosis in this 15-year-old girl? Why was it made?
2. How did a lack of communication contribute to lengthening the time it took to make a correct diagnosis?
3. What did the CT scan reveal in the teenage girl, and how did that result correlate with her abdominal pain?

4. What can be the consequence of pelvic inflammatory disease to the ill patient?

5. In the original diagnosis of appendicitis, why did surgeons prescribe a strong mixture of antibiotics three weeks before the appendix would have been removed?

One cannot overemphasize the need for high immunization coverage levels, especially in children. Indeed, vaccines are the instrument to prevent disease in the population that receives them and protect those who come into contact with unvaccinated individuals within the population. If a child is not vaccinated and is exposed to a disease germ, the child's body may not be strong enough to fight the disease, especially if the child has battled other medical challenges. In this encounter, being unvaccinated left one child dangerously susceptible to an endemic disease.

MYSTERY RASH

CLAIRE PANOSIAN DUNAVAN

Nicolas and Marta Sanchez were worried. For two days, their 10-year-old daughter, Raquel, had been running a fever. Now she complained of irritated eyes, a flushed face, and a cough. Watching her chest rise and fall, her parents counted 22 breaths per minute, a bit faster than normal. This was no ordinary cold.

Raquel's history gave them plenty of reason to worry. She was their miracle child. Ten years earlier, she was born full term with a lusty cry—and with skin and eyes the color of maize. Don't worry, the doctors had said, lots of newborns need a day or two under lights to help clear excess bilirubin—a component of bile—from the blood. Nicolas and Marta had waited patiently. But unlike her nursery-mates, Raquel stayed yellow.

Soon Raquel's doctors squirted dye into her veins in order to scan her liver and gallbladder for signs of disease. Then their looks turned grave. The high level of bilirubin was not caused by broken-down hemoglobin from a mother-infant

Reprinted with permission from the November 2006 issue of Discover *magazine. Copyright © Discover Magazine. All rights reserved. For more information about reprints from* Discover, *contact PARS International Corp. at 212-221-9595.*

blood reaction. The accumulation was a result of malformed liver ducts. Biliary atresia, they called it.

A surgeon tried to create an outlet to divert Raquel's bottled-up bile. Still her color didn't change.

Raquel's last hope was a liver transplant. Shortly after her first birthday, surgeons removed Raquel's liver and replaced it with a healthy one. Eventually, a Y-shaped scar and pills to help her body accept the transplant were the only reminders of her rocky start.

Now Raquel was sick again. Nicolas and Marta knew that antirejection drugs keep transplant patients alive because they paralyze the immune system. What might be a minor flu in another child could be a disaster in Raquel.

Jaime de la Torre, a pediatric infectious diseases consultant, met Raquel Sanchez a few hours after her admission to my hospital. By then, her brown eyes were rimmed with red, and fiery blotches covered her face, arms, and trunk. The intern and resident had already jotted down a differential diagnosis—scarlet fever, toxic shock syndrome, Kawasaki disease, adenovirus, enterovirus, drug rash—and put her on intravenous antibiotics.

De la Torre, a native of Peru, took one look and knew better. "Think again," he said. "What contagious rash still kills nearly a million kids every year?"

The answer was on the tip of their tongue, but they hesitated. That possibility was a disease they—and many of their teachers—had only read about in textbooks. And if Raquel Sanchez really had what De la Torre suspected, where on Earth had she picked it up?

The consultant turned to Raquel. "OK, cara," he crooned, "open wide. Let's see the inside of your mouth." Sure enough, in the flashlight's glare were lacy blue patches opposite her second molars.

"Koplik's spots straight from the textbook. Isolate now!" he barked. "This kid has measles."

Until an effective measles vaccine was licensed in the United States in 1963, measles was one of childhood's greatest perils. And where vaccination rates remain low, it still is. Measles probably causes 10 percent of deaths worldwide in children under age 5.

Measles begins like many other airborne viral infections. First, tiny spheres containing viral protein and RNA breach respiratory linings anywhere from the nose to the lower airways. Then they sweep through the bloodstream, infecting white blood cells. These cells release new spheres that ricochet back to the eyes, ears, nose, throat, and lungs, eventually carpeting every square inch of vulnerable respiratory surface. Now the victim has fever, a hacking cough, and watery eyes. Within hours Koplik's spots—the infection of cells in the mouth—appear, followed by measles' classic rash.

Along with these symptoms come more ominous effects. First, tissues damaged by the measles virus are vulnerable to bacterial infections. That's why pneumonia is the last straw for many measles-infected infants.

But the effects of the virus extend far beyond the lungs. Because the rubeola virus invades a special class of virus-fighting immune cells, called T lymphocytes, it can depress the entire immune system for weeks. Intestinal infection causes the gut lining to slough, tipping many toddlers into malnutrition. And in youngsters whose diets lack enough vitamin A, the shortage of nutrients can swiftly lead to blindness. In rare cases, the infection progresses to the brain, causing neurological damage or death.

Raquel was lucky. She had an ordinary case of measles. Yes, her chest X ray showed a patch of pneumonia in her left lung. But once the antibiotics started to work, her breathing calmed and her cough settled down.

Meanwhile, De la Torre had two more questions for Raquel's parents. Had she ever received a measles shot? And where might she have caught measles? He knew that measles cases in Los Angeles were at an all-time low.

Offhand, Nicolas and Marta didn't remember specific vaccines and years. But Nicolas did recall a recent outing, "*Sí*, two weekends ago, we were in Tijuana. We shopped, ate, listened to mariachis. But we saw no one with spots."

Like most people, Nicolas didn't know that measles is most infectious a day or two before the rash. And because of measles' contagiousness, a single cough or sneeze aimed at Raquel from another viral victim was all that was needed to launch a new infection. The interval from exposure to illness also fit perfectly: 10 to 14 days.

The next day, the final piece of the puzzle fell into place. Raquel's parents brought in her immunization card, and her doctor confirmed that she had never received a measles-mumps-rubella vaccine. Now De la Torre believed he could reconstruct the entire sequence of events.

Not all childhood vaccines are alike. Some shots, such as tetanus, polio, whooping cough, and hepatitis B, are made from killed microbes or their components. However, two vaccines—measles-mumps-rubella and chicken pox—contain live, weakened organisms. In healthy kids, these crippled viruses do no harm, but in transplant patients whose immune systems are held in check by drugs, there's a chance organisms can proliferate out of control.

In the first year and a half of Raquel Sanchez's life, her doctors were simply trying to keep her alive, and the measles vaccination most likely just slipped through the cracks. Her liver transplant kept her alive, but she, like all transplant patients, was now especially vulnerable to infection. What doctor would willfully expose her to an infectious risk, even a weakened vaccine virus? And so her opportunity to be immunized against measles was lost forever.

Of course, there is another scenario. Before her transplant, maybe Raquel's doctors figured that measles had already been conquered at home and that was enough. In other words, she really didn't need the vaccine. If so, they were naive. Today's transplant patients aren't bubble kids. They eat international foods, cross borders, and share air and germs with the rest of the global villagers. Ironically, in doing so, they are sometimes the first to expose chinks in our public health armor.

Before you hop that next flight or plan an exotic getaway, are you sure that you or your children aren't sitting prey for measles?

UPDATE

As this encounter clearly shows, the most important action parents can take to protect their kids from measles and other vaccine-preventable diseases is to have them vaccinated. As a result of a successful immunization program, measles—also called rubeola—was declared in 2000 to be no longer endemic in the United States. Of the 49 confirmed cases of measles reported in 2006, all occurred in under- or nonimmunized individuals who acquired the virus while visiting a measles-endemic country or from such an individual. Globally, the World Health Organization (WHO) estimates that there were 30 million cases of measles in 2005, resulting in approximately 345,000 deaths. This represents a drop of 60 percent since 1999 and is due in large part to the success of the Measles Initiative (supported by the WHO, UNICEF, the American Red Cross, the United Nations Foundation, and the Centers for Disease Control and Prevention). Those countries, especially in Africa, with large measles burdens have now committed to a new goal: to cut global measles deaths by 90 percent of 2000 levels by 2010.

QUESTIONS TO CONSIDER

1. Reconstruct the series of events leading to Raquel's illness.
2. Why were doctors reluctant to administer the measles vaccine to Raquel?
3. Why was Raquel's history of a liver transplant important to understanding her measles infection?
4. How does the measles virus depress the entire immune system for weeks?
5. What is the significance of the development of Koplik's spots in the mouth?

Emergency rooms are often very busy places. Patients can run the whole gamut from seriously ill or having a life-threatening disease to more mundane injuries and wounds. The emergency room physician needs to be ready to tackle any and all cases that come to the emergency room. Sometimes, as in this encounter, what might seem like a common, easily treated illness on first glance can turn out to be more serious than expected.

WHY CAN'T HE WALK?

PAUL AUSTIN

The emergency room was jammed, and 10 patients were waiting to come back from triage. Maria, the charge nurse, stared at the list of patients' names, trying to find a way to open up some beds.

"If some can be seen in the hallway, I'll be glad to start taking care of them," I said. "We can start clearing out the backlog while we open up some spaces."

"Works for me." She asked the triage nurse to bring back a group.

The first was Joey, a blond 15-year-old who'd been having cold symptoms for six days. Good. Should be quick.

"Any other problems?" I asked, turning from the boy to his father.

"Just his runny nose and a sore throat," his dad said.

"Anything else?"

"Nothing. Except he can't walk."

"Can't walk?" Nonsense, I thought. Joey was a trim, muscular teenager.

"He's wobbly," his dad said.

Joey shrugged. "My feet aren't right."

Reprinted with permission from the May 2004 issue of Discover *magazine. Copyright © Discover Magazine. All rights reserved. For more information about reprints from* Discover, *contact PARS International Corp. at 212-221-9595.*

This was looking less and less like a quick fix. "Joey," I said, "touch my finger, then my nose."

I wanted to see if his cerebellum, the part of the brain that handles coordination, was working normally. I expected precise movements from this fit young man, but when he reached out, his finger waggled like a palsied old man's.

"Does he usually have problems with his coordination?"

"No," his dad said. "He's an athlete. Plays basketball."

"Let's see how his walking is," I said. "Joey, take a few steps."

Joey swung his legs off the stretcher and tried to walk. His feet wobbled from side to side, and he had to clutch my arm for support. This was serious.

"You can sit back down," I said, helping him back to the stretcher. "He's had a fever. Any other problems?"

"Well," his dad said, "mostly this cold."

"What about a headache?" I asked the teenager.

"Yeah, kind of."

"Does the light hurt your eyes?"

"Some."

"Here," I said, "let me nod your head for you." I gently moved his head forward and backward. Not really stiff. I was worried about meningitis, a bacterial or viral infection of the membranes surrounding the brain, but if he had that, he'd be sicker by now. Emergency room doctors tend to think of the worst things first, rule those out, and then work back toward less serious causes.

"I need to examine him," I told his dad. "I'm going to move him to a place with some privacy."

I pushed his stretcher past the other patients lining the hallway down to the psychiatric holding room, the only available space. Once we were out of the hall, I did a quick physical exam.

"Your neck looks puffy here." I pointed to the sides of his neck.

"It's always like that," his dad said.

I was doubtful. Necks aren't normally puffy. "Have you been feeling tired lately?" I asked.

"Yeah."

Could it be mononucleosis? That condition results from viral infection, but movement problems aren't common.

"Touch your chin to your chest again." Any stiffness in his neck would support a diagnosis of meningitis. But Joey could move his head without any discomfort.

"Do you do any drugs?"

"Nah, man," he said. "I'm an athlete."

"I didn't think you used drugs," I said, "but I had to ask."

Bacterial meningitis was the only treatable cause I could think of, but it would be unusual to appear this way. And he was too young for a stroke.

"He hasn't hit his head, has he?" I asked his father.

His dad looked at his son. Joey shook his head.

I patted Joey on the shoulder and turned to his father. "This could all be due to a virus, but I'm not sure," I said. "I think we should give him some antibiotics and do a CT scan and a lumbar puncture, just in case it's meningitis. If it is bacterial meningitis, we don't want to fool around."

His dad shrugged. "OK."

I found Maria. "We need to move someone else out into the hallway to make room for the kid with the cold. He can't walk."

Maria gazed at the board. "Room 4 can come out."

Joey was quickly moved to room 4, given a dose of intravenous antibiotics, and then taken for a CT scan. The scan was normal. I did the lumbar puncture; the fluid I removed from his spinal column was crystal clear—no sign of immune cells fighting off an infection.

After I got Joey on antibiotics and started his workup, I called the pediatric residents from a nearby medical school who rotate through our hospital. I wanted them to evaluate Joey. After examining Joey, they arranged a transfer to the school's medical center.

"We're not sure what he's got," one of them said. "If he gets worse, that's the place he should be."

"We've given him antibiotics, in case it turns out to be bacterial." I motioned to the sides of my neck. "But isn't there an encephalitis you can get with mononucleosis?"

"Yeah," the resident said. "I think so."

Encephalitis is an infection of the brain. But Joey didn't have sudden fever, vomiting, and drowsiness—the classic signs of encephalitis. In any case, Joey would be better off under hospital supervision.

I explained the transfer to Joey and his father, and the paramedics came to wheel him away. As soon as he was out of the department, I forgot about Joey. Patients were still waiting for beds.

A week later, I saw the same pediatric resident in the emergency room. "Good call," she said. "EBV encephalitis."

"What?"

"Epstein-Barr virus. Remember the kid with the movement problems? He had EBV, and it spread into the nervous system. It's rare, but it can happen"

"How's he doing?"

"Better," she said. "The resident taking care of him called and told me."

Epstein-Barr virus is a type of herpesvirus, and humans are its only known reservoir. Because the virus spreads primarily through oral secretions, the disease it causes—mononucleosis—is known as the kissing disease. Infection is common: Ninety percent of American adults over age 25 have antibodies to it. Not everyone develops symptoms after infection, however. Those who do develop symptoms are responding to the body's own defense against the virus. Epstein-Barr preferentially infects B cells of the immune system. When that happens in large numbers, the body's defenders against viruses—T cells—go on the attack. The most common symptoms—fever, sore throat, and swollen lymph glands in the neck—result from the T cells' battle against the infected B cells.

The disease leaves many patients feeling exhausted, and it can be weeks before they feel normal. Complications of the infection can include rupture of the spleen, hepatitis, decreased platelet count, anemia, inflammation of the testes, even inflammation of the heart. The virus can also spread into the central nervous system, as in Joey's case. There are no effective medications against the virus, although steroids are sometimes used to treat a few of the symptoms. The good news is that problems resulting from Epstein-Barr are unusual, and when they do occur, they are rarely fatal.

Sometime later I called the medical center to find out how Joey had done.

"I remember him," the resident said. "Joey Jenkins. Nice kid. We were consulted on him. As I remember, the tests indicated the encephalitis was due to EBV; they showed that his antibodies to the virus were from a recent infection."

I heard computer keys clicking over the phone.

"Let me check his discharge summary," she said. "Yup. Hospitalized for two weeks, got physical therapy and a trial of steroids. Improved, discharged to home, still using a walker."

"Will he have a full recovery?"

"Hope so," she said. "He's motivated and has a good chance if he keeps at his physical therapy."

UPDATE

Infectious mononucleosis or glandular fever is often referred to as "mono." The Epstein-Barr virus, which occurs worldwide, spreads through saliva, which is why mono is sometimes called the "kissing disease." However, it also can be spread by coughing, sneezing, or sharing a glass or food utensil. Mono occurs most often in 15- to 17-year-olds, and a blood test can indicate if one has the disease. Most people get better in two to four weeks, and mononucleosis usually isn't very serious, although the virus remains in the body for life. In a few carriers, the virus plays an important role in the emergence of Burkitt's lymphoma and nasopharyngeal carcinoma, two rare cancers that are not normally found in the United States. As seen in this encounter, mononucleosis also can have uncommon complications, including an infection in the nervous system (meningitis, encephalitis, Bell's palsy, or Guillain-Barré syndrome).

QUESTIONS TO CONSIDER

1. In the emergency room, why does a doctor "tend to think of the worst things first, rule these out, and then work back toward less serious causes?"

2. What are the classic signs of encephalitis?

3. Why was Joey put on intravenous antibiotics if he had a viral infection?

4. What are the most common symptoms of an Epstein-Barr virus infection?

5. If Joey had meningitis, how would the fluid from the lumbar puncture have looked?

An infectious-diseases specialist is trained extensively to be knowledgeable in all types of infections, including those caused by viruses, bacteria, fungi, and parasites. Since many infectious diseases, especially protozoal parasites, are strangers in much of the developed world and can be quite unfamiliar to many clinicians, the infectious-diseases specialist plays an important role in correctly diagnosing and treating even rarely seen diseases, as this encounter demonstrates.

BULL'S-EYE

CLAIRE PANOSIAN DUNAVAN

As a child, I was fascinated by a scar on my grandfather's cheek. "How'd you get that?" I asked, touching the shallow depression as we settled into a comfortable chair to read a book.

"Where I grew up, everyone had a scar like that," replied my grandfather, a native of southern Turkey. "First a fly would bite, then you got a sore. After a few months, the skin healed, and you never got the sore again."

Twenty-five years later, as a researcher in Boston, I learned that my grandfather, with no science education, had grasped the biology and immunology of that sore surprisingly well. But I'm getting ahead of my story.

Kabul, Afghanistan, 2000. To escape the heat, Majeeba and Naieda Fazly flopped down on a striped mattress in the large, open-air balcony of the family home they had not seen for almost a decade.

Seven years earlier the Fazly family had suffered a blow. Mr. Fazly, 58, died of a heart attack, leaving behind a widow and 11 daughters. Only one good thing came of the tragedy.

Reprinted with permission from the September 2002 issue of Discover *magazine. Copyright © Discover Magazine. All rights reserved. For more information about reprints from* Discover, *contact PARS International Corp. at 212-221-9595.*

Lacking a male protector in a harshly conservative society, the Fazly women were eligible for international asylum. For several years they waited in Pakistan. Finally, the family was granted refugee status, and they returned to Kabul one last time before leaving forever.

Majeeba and Naieda were the youngest Fazly daughters and mere schoolgirls when they left Kabul for the first time in the early 1990s. Things were bad enough then. But now the streets were filled with trash, mangy dogs, and loud men with beards. And insects. Majeeba and Naieda had never seen so many sand flies in the city. Even in the covered sleeping porch, their bites were incessant.

A year later in Los Angeles, that memory of sand flies was the final clue to their diagnosis. Why? On their hands and forearms, both girls had fleshy, crusted ulcers. Just like the ulcer my grandfather once described.

Leishmaniasis is a protozoan infection transmitted by the bite of a sand fly. Protozoa are single-celled microbes that can develop exotic shapes and accessories: amoebas with crawling pseudopods, ciliates bristling with hairs, and flagellates sporting wavy antennae.

Leishmania assume two shapes during their life cycle. In the first stage, they take the shape of a flagellate that multiplies in the sand fly's gut and eventually swims free in sand fly saliva. In the second stage, found exclusively in humans and animals, they lose their tails and become ovoid bull's-eyes inhabiting immune cells called macrophages. The cycle starts anew when a virgin sand fly siphons parasitized macrophages from a skin ulcer or blood. During the next several days, the ovals morph back into creatures with tails, replicate in the sand fly's gut, and finally migrate to the fly's proboscis, ready to enter their next mammalian host.

The earliest descriptions of skin ulcers containing microscopic bull's-eyes can be found in medical journals from the late 1800s. Then, in 1903, a Scottish army pathologist named Leishman described the same bull's-eye pattern in cells taken from the spleen of a soldier who died after trying to fight off a fever for seven months while stationed at Dum Dum, outside Calcutta. The 23-year-old recruit, Private J. B. of the Second Royal Irish Rifles, became the first-ever reported case of a distinct strain of *Leishmania* that invades blood, bone marrow, and internal organs. Today the disease is known as visceral leishmaniasis. While people can recover without treatment from leishmaniasis that affects only the skin, untreated visceral leishmaniasis is still fatal.

Following Leishman's discovery, names like "dum dum fever" and "kala azar"(both referring to visceral leishmaniasis) and "Baghdad boil," "Biskra button," "Aleppo evil," and "oriental sore" became synonyms for subtypes of Old World leishmaniasis. On the other side of the globe, "uta," "bay sore," and "forest yaws" were linked with New World leishmaniasis of the skin. In time, even the scalloped scar that followed leishmaniasis of the outer ear acquired its own nickname. Because this ragged defect was common in Latin American men who harvested chicle for making chewing gum, it was called "chiclero's ulcer."

One reason that the form affecting the skin, cutaneous leishmaniasis, thrives in so many locales worldwide is its large range of reservoir hosts. In Central and South America, rodents, marsupials, sloths, and anteaters carry local species such as *L. mexicana*, *L. brasiliensis*, *L. panamensis*, and *L. guyanensis*. In Old World urban settings like Kabul, dogs often harbor *L. tropica*. And in less populous areas of the Old World, rats and gerbils are commonly infected with *L. major*. In dry, desert areas of central Asia, Iran, and Iraq, up to 30 percent of gerbils have ulcerative skin lesions of *L. major* affecting their head, ears, and the base of the tail.

Fortunately, localized infections of *L. tropica*—the canine cutaneous strain that infected my grandfather as well as Majeeba and Naieda—often heal without specific treatment, although it may take many months. As a young researcher, I tried to understand why. I did so by inoculating lab mice with *Leishmania* and observing their skin lesions. Depending on the genetic and immunologic makeup of the mouse, the sore would either mushroom or cure itself over time. Other investigators went further, pinpointing specific lymphocytes and products they released that governed healing in mice and humans. Just as my grandfather said, once bitten and healed, you were virtually always protected against repeat infection by the same strain.

But Majeeba and Naieda had a different dilemma. One year had passed, and they still hadn't healed their primary infections. Their diagnosis wasn't in question. Aside from the evidence staring me in the face, skin biopsies sent by their referring dermatologist showed unmistakable bull's-eyes. Could they have an underlying illness or immune deficiency to explain their persistent infection? My clinical instincts said no. Sure enough, after testing, the sisters proved perfectly healthy, merely frustrated by their ugly nodules and scabs rimmed with scarlet.

In the end, I never did figure out why their ulcers were chronic. It could have been the stress of moving or perhaps repeated infection through multiple bites. No matter what the reason, the girls needed treatment. The good news was that we had options. When I was doing bench research on leishmaniasis in the 1980s, a 20-day series of daily injections of intramuscular or intravenous antimony, a metal that has long been used as a medical treatment, was standard first-line therapy. In 2001, I chose instead itraconazole, a modern drug taken by mouth that often speeds healing in mild to moderate cases of leishmaniasis that affect the skin.

My gamble paid off. Months later the ugly scabs were flat, mauve stains, well on their way to disappearing completely. More important, the two sisters from Kabul were making a happy adjustment to their new life in America. Majeeba was working at a local department store, and Naieda was finishing high school.

Throughout history, war and civil strife have often played a role in outbreaks of leishmaniasis. One recent example is the devastating epidemic of visceral leishmaniasis that began in 1988 in the western Upper Nile province of southern Sudan. Fueled by famine and a brutal government-led campaign against civilians, it has claimed more than 100,000 lives. Cutaneous leishmaniasis, although less dangerous, is still a problem for military personnel in areas like the Middle East and Panama. And an epidemic of leishmaniasis is raging in Afghanistan; no doubt a few troops will return from the region bearing crusts and scars of infection.

Meanwhile, leishmaniasis continues to surprise researchers. For example, a handful of U.S. soldiers who served in the Gulf War came down with visceral infection due to *L. tropica*, a species previously confined to the skin. This unprecedented event triggered a temporary ban on blood and organ donations by all Desert Storm veterans. Visceral leishmaniasis has also emerged as an opportunistic infection among patients with HIV in Spain, Italy, and southern

France. And, in rodents and test tubes, leishmaniasis is still an ideal model system to study immune responses.

Would any of this have interested my grandfather? Probably not. My guess is that as far as he was concerned leishmaniasis was simply one more thing life dished out, like the common cold or the irksome bite of a sand fly.

UPDATE

Leishmaniasis is found in parts of about 88 countries where, according to the World Health Organization (WHO), approximately 350 million are threatened, 12 million are infected, and 2 million new infections occur every year. More than 90 percent of the world's cases of visceral leishmaniasis are in India, Bangladesh, Nepal, Sudan, and Brazil. Visceral leishmaniasis is a serious but preventable health threat to deployed U.S. Service members in Iraq and Afghanistan. Military officials have estimated that there have been more than 2,500 cutaneous infections in service members between 2003 and 2007. About 95 percent of these acquired cases are self-healing infections caused by Leishmania major. *Although it has been rare for someone to contract the disease in the United States, more cases of this infection are now appearing. In 2007, individuals in North Texas, who have not traveled to endemic areas, are contracting the disease. Culturing of the parasite from these patients has identified it as* L. mexicana, *a less dangerous form of the parasite that appears to reside in wood rats and is transmitted by a sand fly.*

QUESTIONS TO CONSIDER

1. Describe the two stages of leishmaniasis.
2. Why is one form of the disease called visceral leishmaniasis?
3. What were the two reasons suggested by the infectious diseases specialist as to why Majeeba and Naieda had chronic ulcers?
4. How is leishmaniasis treated?
5. How did Majeeba and Naieda get infected with the parasite?

Often sexually transmitted diseases (STDs) are not apparent to the person infected, making it unlikely that the individual would go to a physician or clinic for treatment. If the STD then develops into a more severe and dangerous infection, it might be too late for the physician to institute procedures to save the patient. As this encounter relates, survival can have serious consequences for the patient's future.

CAN SHE SURVIVE THE CURE?

STEWART MASSAD

A monitor in the intensive care unit at Cook County Hospital in Chicago continuously beeped and flashed alarms as the young woman's pulse rose above 140 beats per minute and her blood pressure fell. Her skin was ashen and clammy, her muscles slack, her lips cracked, her eyes rolled up under limp lids. It was clear she was dying. What wasn't clear was whether she could withstand the extreme measures required to save her.

The patient was in her late twenties and had begun experiencing severe pelvic pain following her last menstrual period two weeks before. By the time her mother brought her to the emergency room, she couldn't stand up straight because of swelling from an anaerobic bacterial infection in her belly. A DNA probe identified the original infection as *Neisseria gonorrhoeae*, which attaches to cells in the reproductive tract that nourish sperm. Although cervical mucus contains antibacterial compounds, menstrual blood can wash the mucus out, making it easier for *Neisseria* and other microbes to migrate up the reproductive tract and into the fallopian tubes. Her gynecologists put her on intravenous antibiotics and waited

Reprinted with permission from the June 2005 issue of Discover *magazine. Copyright © Discover Magazine. All rights reserved. For more information about reprints from* Discover, *contact PARS International Corp. at 212-221-9595.*

for her to heal. Instead, after four days, her infection had only grown worse.

A century ago, pelvic infections from gonorrhea were a major source of disability and one of the causes of the "female complaint" many patent medicines were marketed to relieve. In that preantibiotic era, many women suffered from lingering infections that could sometimes kill. Even today, about 150 women in the United States die each year from complications of pelvic inflammatory disease. This patient was about to become one of them.

As gonococcal bacteria multiply in the fallopian tubes, components in the bacterial cell walls rouse the immune system defenses. Antibodies bind to the cells, marking them for engulfment and destruction by white blood cells. Substances in the blood called complement proteins punch lethal holes in the bacteria. There is also a structural barrier: Loops of bowel and a membrane called the omentum stick together, shielding the abdominal cavity from bacterial infiltration. But in a few cases, bacteria get through. The fallopian tubes fill with pus, an acidic soup of toxic enzymes and exhausted immune cells. As the pus collects, the body walls it off in a capsule formed of a blood component called fibrin. That pocket, or abscess, contains the infection. An abscess lacks blood vessels, so white blood cells, substances secreted by immune cells, and antibiotics have difficulty reaching the infection. The body tries to make up for the lack of blood flow: Small blood vessels grow leaky to permit immune defenses to seep out into infected tissue. But the leakiness also allows bacterial products into the bloodstream, which can further ignite an inflammatory response. Eventually, the body cannot keep the arteries filled, and blood pressure falls. Heart function falters. Deprived of blood flow, the major organs fail, and the patient dies.

The only cure is to operate—drain the pus and remove the dead tissue. But inducing anesthesia in a critically ill patient is risky. The drugs that bring on sleep can depress the function of an already compromised heart, and anesthesia-induced paralysis curtails the function of the lungs. And if the clotting factors in the blood have been used up, the patient can bleed to death. So doctors often put off surgery until the crisis is unmistakable: The abdomen is rigid, the fever uncontrollable, the patient delirious and writhing in pain.

I am a gynecologic surgeon, and when this young woman was clearly in crisis, her doctors called me. The staff loaded her with enough fresh-frozen plasma to replace lost clotting factors. Then, after we moved to the surgical theater and the anesthesiologist completed the sedation, I quickly made an incision from navel to pubic bone. The pus spilled out. We suctioned the abdominal cavity and looked for the source of the infection.

The wall of one fallopian tube was dead, ruptured, the yellow-brown pus escaping under pressure through a red-black hole. The other tube, scarlet from inflammation instead of a healthy pink, was minimally involved. We removed the damaged fallopian tube. Then we cut the abscess away from the places where its fibrin walls clung to the fallopian tubes, to the colon and small bowel, to arteries and veins, and to the uterus. More pockets of pus lay hidden in the cavity between the cervix and the rectum, in the trough along the descending colon, and under the ovary.

We broke the pockets of pus open and cleaned the area. A generation ago, it was the custom to remove the uterus once an infection had advanced into the fallopian tubes. Now newer intravenous antibiotics can sterilize moderate infections after the abscess is drained. Finally, we peeled the remaining walls of the abscesses away from the peritoneum, the membrane that lines the abdomen and pelvis. It was like removing the rind from an orange. We sent samples to the lab to make sure the infecting microbes were sensitive to the antibiotic treatment. We checked for bleeding, washed the abdominal cavity with warm saline solution, and closed the incision.

The day after the operation was touch and go. The patient needed a tremendous amount of fluid to keep her kidneys, brain, and heart working. She required blood transfusions to deliver oxygen to these organs and to the site of the infection. She lay sedated on a ventilator in the intensive care unit all that day and all the next, and then she rallied. Her blood pressure and urine output rose, her pulse fell to normal, and her fever began to subside. The staff tapered her off sedatives, and she woke up. After a terrible vigil in a tiny waiting room, her mother went home. After a week, the young woman followed.

Each year about a million American women are treated for pelvic inflammatory disease, and sexually transmitted infections are usually the cause. The longer the infection remains untreated, the greater the risk of infertility. This patient's infection had advanced, and only time will tell whether she will be able to conceive. If scar tissue does not obstruct her unaffected fallopian tube, she might one day have children.

UPDATE

Pelvic inflammatory disease (PID) is an infection in the female reproductive organs (uterus, fallopian tubes, and ovaries). In 2007, it was the most common preventable cause of infertility in the United States, where more than one million women are affected by PID, and some 50,000 become infertile each year. PID usually results from a sexually transmitted disease (STD) that hasn't been treated. Thus, Neisseria gonorrhoeae *and* Chlamydia trachomatis—*two bacterial species associated with STDs—are the most common causes of PID, but other bacterial species also can be involved. If a person is infected but untreated, it can take from a few days to a few months to develop into PID. Antibiotics can cure PID, so early treatment is important. If PID is not treated, or is treated late as in this encounter, it can lead to severe problems including infertility, ectopic pregnancy, constant pelvic pain, and abscess formation. If a woman is treated for PID caused by an STD, it is important that her partner also be treated. Unless the partner is treated, the infection may reoccur and cause even more damage.*

QUESTIONS TO CONSIDER

1. What is the body attempting to do by walling off the pus in an abscess?

2. What factors make it more likely that an organism, like *Neisseria gonorrhoeae*, can migrate up the female reproductive tract and into the fallopian tubes?

3. What is the cure for cases of advanced pelvic inflammatory disease where abscesses exist?

4. Why do physicians often postpone surgery for advanced cases of pelvic inflammatory disease until the symptoms are unmistakable?

5. What are the unmistakable signs and symptoms of pelvic inflammatory disease?

Infections of the gastrointestinal system do not always suggest to the physician an obvious cause. Appendicitis may be the culprit, but when the signs and symptoms do not completely match up, another diagnosis must be developed. As this encounter relates, sometimes several medical diagnoses by a variety of medical experts must be considered before the physician can arrive at the final verdict.

GUT ATTACK!

TONY DAJER

The patient's sister pointed to the form lying on the stretcher. "It's pouring out of him, doctor." Then she lifted the sheet. "We even had to put these on."

The young man, looking gaunt and miserable, reflexively pulled up his knees for cover. But there they were, diapers as puffy as pantaloons. I stared, trying to remember if I'd ever seen Pampers on a 22-year-old.

"How often?" I asked.

"All the time," the sister said. "It never stops."

I restudied the chart: His blood pressure was normal at 120/76. He didn't have a fever, but his heart rate was 150. His lips were cracked, his eyes sunken, his skin shriveled—this patient was as dry as the Mojave. "When did it start?" I asked.

"Two days ago, doctor," the patient rasped.

"Vomiting?"

"Sometimes."

"Have you traveled?"

"No."

"Anybody else sick at home? Any pets?"

"No. No."

Reprinted with permission from the July 2005 issue of Discover *magazine. Copyright © Discover Magazine. All rights reserved. For more information about reprints from* Discover, *contact PARS International Corp. at 212-221-9595.*

"Have you had pain?"

"Everywhere."

I felt his belly. No hot spot of tenderness—just soreness all over. "And which came first?" I continued. "The vomiting or the cramping?"

He pulled the sheet a little higher. "The vomiting. Then the water coming out." He winced. "Then so much pain."

The diagnostic beads clacked together: vomiting first (appendicitis usually causes pain first) plus soft belly plus lots of diarrhea plus no fever. The answer: gastroenteritis (*gastro* = stomach, *enter* = intestine, *itis* = inflammation, a doctor's job being, of course, to tell you in Latin what you just told him in English).

The emergency room nurse asked if I wanted to get more blood tests.

I was about to decline—the average stomach flu rarely throws an otherwise healthy young man's electrolytes out of whack. But then I remembered his 150 heart rate. And the Pampers. "Sure. Thanks," I said.

Gastroenteritis, stomach flu, food poisoning. Few maladies beget more misery, put the kibosh on more vacations, or make the food industry jump through more hoops than intestinal infections. In ascending order of size, the culprits include viruses, bacteria, and protozoa. A viral infection, for instance, set off cruise ship outbreaks in 2002; in 1993 a protozoan that spread through the water filtration system sickened 400,000 people in Milwaukee—almost two-thirds of the city's population.

Bacteria, for their part, can provoke loss of lunch or loosening of bowel from the toxins they produce or by direct invasion or both. *Staphylococcus*, for example, is almost ubiquitous on people's skin. Spread onto food at room temperature, one strain of the bacterium, *Staphylococcus aureus*, can multiply and secrete a heat-resistant, vomiting-inducing toxin. Other pathogens, such as cholera-causing *Vibrio cholerae* and *Escherichia coli*, must first get into the intestine before they can do their damage. Once in the gut, *V. cholerae* secretes a toxin that provokes the human intestine to shed up to one liter of fluid per hour, thus facilitating the pathogen's spread to other hosts. The invasive *E. coli* takes a different tack: It attaches to cells in the intestine, then secretes a toxin that provokes cramping and diarrhea. The immune response to the infection provokes a bacteria-killing fever. One sign of the widespread inflammatory response is white blood cells in the stool.

Areas with poor water sanitation—mostly in the developing world—are at greatest risk for these bacterial infections. But the United States is hardly exempt: The *Salmonella* family alone—proud sponsor of over 2,400 disease-producing strains—is estimated to have caused 2 million cases of bacterial gastroenteritis last year. Although the infection rarely progresses very far, it can be lethal in those with compromised immunity: the very young, the old, or the debilitated.

Two hours later, the young man's lab results surprised me—and renewed my respect for these germs. His white blood cell count was normal, but the level of urea nitrogen, a measure of dehydration, was sky-high. Even more worrisome, the creatinine level—a measure of kidney function—was twice as high as it should have been.

"Wow, he's even drier than I thought," I said to the nurse.

"On his third liter," she replied. "That should be helping. But I just retook his temp. It's 101.8."

"I don't have the white-cell stool smear back yet," I said. "But he's sick enough to deserve antibiotics. This must be an invasive enteritis."

"What do you want?"

"Cipro, 500 milligrams." Ciprofloxacin acts against a broad spectrum of bacteria.

"Coming up."

When I went in to check on the patient, his heart rate was down to 105, but he still looked like a rag doll.

"I have no energy, doctor."

I pointed to the IV bag. "This will help you." His sister joined me at his bedside. "He needs to be admitted," I told her. "We're going to keep up the fluids and start some antibiotics."

She caressed her brother's forehead. "He has been so sick."

The most likely culprit was *Salmonella*, which prospers in the conditions of modern food processing. Most other bacteria that cause foodborne illnesses stick to a preferred niche: lesser-known *Campylobacter* is common in chickens, and *Aeromonas* often crops up in milk; *Shigella* trots around via the water supply or dirty toilets; *Yersinia enterocolitica* prefers pigs; *Vibrio parahaemolyticus* infests uncooked shellfish, especially oysters. *Salmonella*, on the other hand, is promiscuous.

It inhabits chickens, ducks, turkeys, dogs, and unpasteurized milk; turtles, iguanas, and even rattlesnake meat have caused outbreaks. By some estimates, 80 percent of *Salmonella* infections come from raw or undercooked eggs. Chickens not only spread it to each other but also contract it from inexhaustible reservoirs of rodent feces and farm waste. In 1994 one outbreak of 224,000 cases was tied to an ice-cream mix hauled in a container truck whose previous cargo had been unpasteurized eggs.

The good news is that thorough cooking (no runny yolks) destroys *Salmonella*, and continuous refrigeration thwarts it. The bad news: You can't cook your dog. Moreover, the overuse of antibiotics on farms, along with injudicious use of prescription antibiotics among people, leads to more and more drug-resistant bacteria.

Using antibiotics against *Salmonella* is tricky because drugs can prolong bacterial shedding from the GI tract and increase the chance of spreading the infection. A mild infection can be left to run its course, but severe cases must be treated. The difficulty is that stool cultures can take 48 hours or more to confirm the diagnosis, leading to unhappy shoot-first, answer-questions-later situations.

After two hours, the nurse grabbed me. "His belly pain is worse."

I walked over. The patient was clutching his middle. "Where does it hurt?"

"Here," he groaned, waving a hand across his lower abdomen.

I pressed all around. He seemed most tender on the right: Appendicitis. "I've been sitting on an appie," I muttered to the nurse. "He needs a CT scan."

This curveball shouldn't have surprised me. Some invasive bacteria, like *Y. enterocolitica*, mimic appendicitis so perfectly that outbreaks come to light only after doctors notice a spate of negative appendectomies. And more than 100 years ago, physician William Osler singled out typhoid fever—caused by *Salmonella typhi*—as one of the most protean and deceptive of diseases; before his students could call themselves doctors, he made them master its diagnosis.

The CT scan would take a while—an hour for the patient to drink the oral contrast, well over an hour for it to percolate down, and a half hour for the scan.

A seeming eternity later, the radiologist called: "Intussusception."

"You're kidding," I blurted out, then kicked myself for the hours I'd wasted. Intussusception can be a surgical emergency; it meant that one part of the patient's intestine had telescoped into another.

The surgical resident on the case was equally surprised. "No way," he said, after examining the patient. "I'm going to talk to the radiologist."

Twenty minutes later he was back with a revised diagnosis. "Colitis," he declared. "Actually, pancolitis. Inflammation all over the colon. You really have to stare at the scan."

Later that evening, the stool smear result came back: "sheets of white cells seen." Although white blood cells can be a sign of a noninfectious condition such as ulcerative colitis, they were more likely a sign of widespread infection.

When I saw the patient two days later, he looked much better. I ran into the surgical resident in the hallway. "Looks like the stool culture is growing *Salmonella*," he said. "Still don't know the strain."

"No pets, no travel, no turtles," I pointed out. "Where did it come from?"

"Beats me," he said. "But I did tell him, 'Wash your hands, keep food in the fridge, and make sure you really fry those eggs.'"

UPDATE

Every year, approximately 40,000 cases of Salmonella *infections are reported in the United States. Because many milder cases are not diagnosed or reported, the actual number of infections may be more than 1 million per year. The bacteria reside in the intestinal tracts of animals, including birds, and can be transmitted to humans by eating foods contaminated with animal feces.* Salmonella *may also be found in the feces of some pets, especially those with diarrhea, and people can become infected if they do not wash their hands after contact with these feces. Determining that* Salmonella *is the cause of an illness depends on laboratory tests that identify the bacteria in the stools of an infected person. These tests are sometimes not performed unless the laboratory is instructed specifically to look for the organism.* Salmonella *infections usually resolve in 5 to 7 days. As this encounter relates, however, sometimes a patient may become severely dehydrated. Between January 1, 2007 and October 12, 2007, at least 174 isolates of a* Salmonella *strain were collected from ill persons in 32 states. When the Centers for Disease Control and Prevention (CDC) undertook an investigation comparing foods eaten by ill and healthy persons, the investigation indicated that eating Banquet brand frozen chicken or turkey pot pies was the likely source of the illness. Because foods of animal origin may be contaminated with* Salmonella, *people should not eat raw or undercooked eggs, meat, or poultry, or drink raw (unpasteurized) milk. The frozen pot pies were recalled.*

QUESTIONS TO CONSIDER

1. What were the signs and symptoms that indicated gastroenteritis to the physician?
2. What was the lab result that suggested the patient was suffering from pancolitis?
3. What are some of the bacterial species and the toxins they produce that can lead to gastroenteritis?
4. What types of "culprits" can cause gastroenteritis?
5. Why was *Salmonella* the likely "culprit" of the patient's gastroenteritis?

There are few infectious diseases that are almost always fatal once they are contracted and symptoms appear. However, if caught early, antisera and vaccinations may provide a positive outcome. Unfortunately, these interventions are not available to everyone. This encounter draws attention to one of the most deadly infectious diseases and the tragic outcome that often results.

A KILLER RAVES ON

CLAIRE PANOSIAN DUNAVAN

In a darkened room, a 9-year-old boy lay on a mattress, loosely bound with a cloth rope. Occasionally, he jerked or twitched. Mostly he stared. His breathing was irregular. He was dying.

One week earlier, the boy had arrived fighting and gagging. For several days, his family grieved at his side as he writhed and choked at the mere sight of water. Now that he was so close to death, his parents had left to arrange the funeral.

I witnessed this case of rabies in 1987, when I was a visiting professor of infectious disease at the Aga Khan Medical School in Karachi, Pakistan. The main hospital, with its brick buildings and leafy courtyards, was surprisingly like my own in Los Angeles. But on this day I had traveled to a public hospital on Karachi's outskirts. The facility housed patients with diseases that American doctors had only read about. My host, Dr. Mohammed, had told me the boy's story and invited me to see him.

At first, the child's condition so unnerved me that I had to look away. Still, a voice inside me said, "Look and remember." While I gazed at him, my thoughts turned to a colleague

Reprinted with permission from the March 2003 issue of Discover *magazine. Copyright © Discover Magazine. All rights reserved. For more information about reprints from* Discover, *contact PARS International Corp. at 212-221-9595.*

who had visited Mexico a few years earlier. During his stay in a remote village, Matt had noticed a dog behaving oddly. One evening, from out of the darkness, the dog snarled and lunged at his leg. When he undressed, my friend found bloody marks where teeth had punctured his calf.

"I instantly knew I had three options," he later told me. "In my halting Spanish, I could try to find the dog's owner and inquire if the dog had ever been vaccinated. Or I could ask to have the dog killed and his brain examined. Or I could get a series of shots." Two days later, Matt was in San Diego receiving rabies antiserum and the first of five rabies vaccinations. Fortunately for him, a new vaccine for rabies had just been approved in the United States. For Americans, the days of painful injections of rabies vaccine in the stomach following unprovoked animal bites were over.

Halfway around the globe, of course, pain was not the issue. For many poor people in India and Pakistan—where thousands of rabies exposures occur every year—treatment was either unaffordable or unavailable. By contrast, improved treatments have helped bring deaths from rabies in the United States down from more than 100 a year a century ago to about one or two every year.

When someone is treated for rabies infection, the goal is to arrest the deadly virus before it reaches the spinal cord or the brain. The pathogen, which commonly gains entry via saliva from a rabies-infected animal, breeds first in local muscle, then advances through long, lanky nerve cells. An initial injection of rabies antiserum (a highly specific antibody culled from rabies-immune humans) can be thought of as a stun gun that slows the virus. Meanwhile, five doses of rabies vaccine given over four weeks are the bullets that complete the counterattack. They do so by kindling enough native antibody to wipe out the remaining invaders.

Before treatment became available, rabies was one of the most uniformly fatal viral infections. As early as the 23rd century B.C., the legal code of the Babylonian city of Eshnunna refers to a disease that was probably rabies. In 500 B.C., the Greek philosopher Democritus recorded an unmistakable description of canine rabies. The word *rabies* itself comes from the Latin verb *rabere*, "to rave," as well as a Sanskrit word for doing violence, underscoring a frequent but not universal feature of the virus's deadly assault.

The two common forms of rabies are "furious" and paralytic, or "dumb." Furious symptoms, such as hydrophobia, delirium, and agitation, reflect invasion of the brain by rabies virus. But in one out of five cases, the disease seems to target only the spinal cord and brain stem. These victims experience confusion and weakness but not the wild, explosive behavior that still prompts straitjackets and padlocked cells for victims in some parts of the world.

In 1892 the renowned physician William Osler described hydrophobia in his medical textbook. "Any attempt to take water," he observed, "is followed by an intensely painful spasm of the muscles of the larynx and the elevators of the hyoid bone [a horseshoe-shaped bone situated at the base of the tongue]. It is this which makes the patient come to dread the very sight of water. . . . These spasmodic attacks may be associated with maniacal symptoms. In the intervals between them the patient is quiet and the mind unclouded."

Osler's text goes on to state that rabies victims rarely injure attendants during their violent episodes, although they may "give utterance to odd sounds." Mercifully, hydrophobia usually gives way to deeper unconsciousness within three or four days. Soon after, organs fail and the heart stops.

Leaving the ward, Dr. Mohammed and I headed for a nearby lounge where we could talk and drink tea. "The boy liked dogs," Dr. Mohammed said quietly as we walked. "It's almost certain he was infected by a dog." I had seen many street dogs in the slums of Karachi, including a few that were sick and whimpering. How easy it would have been, in an impulsive gesture, to reach out to one of them.

Of course, many other animals contract and transmit rabies. Since the 1980s, the silver-haired bat and its kin, the eastern pipistrelle, have been the source of roughly two-thirds of all human rabies cases in the United States. Other susceptible mammals include wolves, foxes, coyotes, cats, skunks, raccoons, and even horses and livestock.

Most human victims sicken within three months of exposure to rabies, but sometimes infections remain dormant for a year or more. Once symptoms begin, however, the die is cast. Over the next one to three weeks, the condemned patient sinks as the virus relentlessly moves from muscle to nerve to spinal cord or brain.

Ancient healers espoused an array of immediate postexposure antidotes, from caustics and cupping to applying a poultice of goose grease and honey. In the first century A.D., Celsus, a Roman physician-naturalist, recognized that saliva transmitted rabies and recommended sucking or burning suspect wounds. Eighteen hundred years later, William Osler's suggestions were surprisingly similar: careful washing, chemical cauterization, and keeping the wound open for several weeks.

Osler did not know of Louis Pasteur's landmark research, conducted just a few years earlier. Pasteur had reported experiments leading to the world's first animal rabies vaccine. He wasn't planning to use the vaccine on a human until he learned of a desperate case: a 9-year-old boy from Alsace bitten 14 times by a rabid dog. Of his next action, Pasteur wrote: "The death of this child seemed inevitable, and I decided, not without lively and cruel doubts . . . to try in Joseph Meister the method which has been successful in dogs. Consequently, on July 6 at 8 in the evening, 60 hours after the bites, in the presence of Doctors Vulpian and Grancher, we inoculated under a skin fold in the right hypochondrium [upper abdomen] of the little Meister a half syringe of the [spinal] cord of a rabid rabbit preserved in a flask of dry air for 15 days." After 12 more injections, it was time to watch and wait. Joseph Meister never developed rabies.

Two 9-year-olds—one in modern Pakistan, one in 19th-century France. One was a rabies victim, the other a rabies survivor. Today an updated form of Pasteur's remedy saves countless lives, yet every year 40,000 to 70,000 people die for the lack of it. If the 19th century's greatest microbiologist knew of these ongoing 21st-century tragedies, what would he say?

UPDATE

In 2005, the Centers for Disease Control and Prevention (CDC) reported two human cases of rabies in the United States. As this encounter describes, without treatment, rabies infections are usually deadly. In fact, globally, rabies causes one human death every 10 minutes and the World Health Organization (WHO) estimates there are some 10 million people who receive post-exposure treatments each year after being exposed to rabies-suspect animals. However, in 2004, the first documented case of a person surviving rabies following the development of symptoms was reported in a 15-year-old girl in Wisconsin. With parental consent, her eight physicians tried a never-tested method: they used drugs to induce a coma state and let the teenager's body fight the virus. Remarkably, she survived and was brought out of the coma after 7 days. Still, the girl has had to go through extensive physical therapy to regain her lost physical functions. The teenager was bitten by a rabid bat, which was just one of more than 5,900 cases of animal rabies reported by the CDC in the United States. More than 90 percent of the animal rabies cases occurred in wild animals found in different geographic areas: raccoons (eastern United States); skunks (north and south central United States and California); bats (in all states except Hawaii); foxes (Alaska, Arizona, and Texas); and mongoose (Puerto Rico). Raccoons are the most common wild animal with rabies, although most human cases originate from a bite by an infected bat. The number of rabies cases in wild animals has been declining in recent years probably due to oral vaccination of wildlife species. Reported cases of rabies in domestic animals, such as cats and dogs, remain low primarily because of high vaccination rates.

QUESTIONS TO CONSIDER

1. Why would a bite on the leg from a rabies-infected animal be somewhat less urgent than a bite on the face?

2. How do the "furious" and paralytic forms of the disease differ?

3. What is the pathway followed by the rabies virus once symptoms begin?

4. Why do many rabies patients experience a fear of water?

5. Identify the types of animals that can transmit the rabies virus.

Life for an emergency room physician or nurse can be fast paced and can require quick decisions to save a person's life. As this encounter relates, fast action can apply to some infectious diseases as well, where a few hours could mean the difference between life and death.

WHY ARE HIS EYES CROSSED?

Mark Cohen

The call came after I had drifted off to sleep, late on a Thursday night. "Hi, Mark, it's Pete Parsons in the ER. I have an eight-month-old here I'd like you to take a look at. He has a fever and the mom tells me that he's been real fussy all day. What puzzles me is that he has a funny kind of strabismus, and the mom is worried. She says his eyes didn't look like that before."

Suddenly I was wide awake. A seasoned emergency room physician who handles heart attacks and major trauma with calm competence can get very nervous when faced with a sick infant. I expect that reaction. But sometimes there are red flags that tell me I need to rush to the hospital immediately. This was a red flag: The baby had strabismus, or crossed eyes. Normally, that's not an emergency, but the mother said the boy's eyes hadn't been crossed before. And she said he had fever and fussiness. That combination of symptoms could mean trouble.

When I arrived at the emergency room, the nurse directed me to the curtained-off enclosure where little Jesse Rivera lay on a gurney. He was awake, but he wasn't moving much. And he wasn't looking at his mother, who was leaning anxiously

Reprinted with permission from the April 2003 issue of Discover *magazine. Copyright © Discover Magazine. All rights reserved. For more information about reprints from* Discover, *contact PARS International Corp. at 212-221-9595.*

over him, stroking his small hand. "Uh-oh, this could be a sick kid," I thought. To a pediatrician, "sick" means "I'd better do something, or this baby might die."

Jesse's cry wasn't the loud, vigorous sound of a baby who is frightened or angry or in pain. His cries came in short, repetitive bursts, with an unusually high pitch. I hadn't heard cries like that in years, but it's a sound you never forget. This was a sick baby.

When I looked at Jesse's eyes, I could see that they were crossed. I could also see that this was not typical strabismus, a condition in which one eye points slightly inward or outward while the child looks straight ahead. I caught Jesse's attention with one of the brightly colored stickers I use to test infants' vision. When the sticker was directly in front of him, his eyes were straight. But when I moved it to either side, his eyes quickly became crossed. I saw the problem: Although both of Jesse's eyes could move toward the center, neither could move to the side (toward his temples). When he tried to look to one side, the eye that appeared crossed actually moved correctly while the other one remained stuck looking straight ahead.

Something was interfering with the signaling in some of the nerves that controlled Jesse's eyes. Twelve pairs of cranial nerves run from the brain stem to various areas of the head and neck. Working together, three of these nerve pairs—the third (oculomotor), fourth (trochlear), and sixth (abducens)—control the muscles that allow your eyes to move together in tiny, precise jumps across this page, leap quickly up to watch the bird on the windowsill, and return just as quickly to the page.

I suspected that Jesse's problem lay in the sixth pair of nerves. These nerves run around the outside of the brain stem, a location that makes them vulnerable to compression when pressure within the skull increases. Bleeding, tumor, trauma, or infections are all conditions that can increase intracranial pressure. The additional pressure sometimes pinches the sixth nerve—on one or both sides—thus paralyzing the muscle that controls the eye's lateral movement.

I was fairly certain that Jesse's malfunctioning nerves were the result of meningitis, an infection of the meninges, which are the membranes that surround the brain and spinal cord. He had a number of classic signs of the disease: fever, listlessness, the sixth-nerve palsy, and that high-pitched, irritable, meningitic cry, also probably due to increased intracranial pressure. Jesse needed immediate treatment if he was to avoid the worst complications of meningitis: seizures, coma, and death.

In most cases, meningitis occurs when infecting organisms—viruses or bacteria—circulate through the bloodstream and lodge in the membranes that cover the brain and spinal cord. Jesse might have picked up the virus from a playmate (in whom it may have caused only a mild cold). Or the cause might be a bacterium, one that typically lives harmlessly in the nose or throat until it suddenly gains access to the bloodstream during a minor illness. The wayward organisms then manage to find an inviting home in the cerebrospinal fluid, the watery layer that protects and cushions the brain and spinal cord.

With a virus, the infection is usually relatively mild and self-limited, though serious aftereffects—such as hearing loss—can occur. A bacterial infection, on the other hand, generally causes progressive inflammation of the meninges, swelling of the brain, and, if untreated, death or serious permanent brain damage. This can happen in a hurry. I didn't know how much time Jesse had before his brain would sustain irreversible damage. His crossed eyes told me he was already showing signs of brain swelling. I knew that I was dealing with minutes or hours, not days.

Usually I make the diagnosis of meningitis by taking a sample of cerebrospinal fluid and sending it to the lab to see if white blood cells or bacteria are present. The lab will also culture the sample to see if bacteria or viruses are growing. I collect the fluid for the sample by performing a lumbar puncture, also known as a spinal tap. I slide a needle between the vertebrae into the spinal canal and remove about a half-teaspoonful of fluid. The needle enters the canal well below where the spinal cord itself ends and the procedure is generally very safe. In Jesse's case, however, removing cerebrospinal fluid from the spinal canal was potentially catastrophic. The spinal tap could allow the increased pressure inside his skull to push the lower part of his brain a few millimeters down through the base of the skull, which might completely stop his breathing.

Instead, I made what is called a clinical diagnosis of meningitis, meaning one based on my history and physical examination without the benefit of laboratory tests. I immediately gave Jesse high intravenous doses of the antibiotics cefotaxime and vancomycin to target all the bacteria that were most likely to be causing his infection. I also added a powerful steroid, dexamethasone, to reduce the inflammation of the meninges and try to prevent further damage to his brain. I called the children's hospital, and they sent an intensive care team to pick him up in an ambulance.

Fortunately, Jesse responded really well to the treatment. A spinal tap, done carefully in the intensive care unit once he was stabilized, showed that his meningitis was caused by the bacterium *Streptococcus pneumoniae*, also called pneumococcus. This organism frequently causes pneumonia in children and adults, hence its name. (A less virulent form of this bacterium commonly causes ear infections in children.) He was placed on intravenous penicillin and gradually regained his normal level of activity and alertness, though he remained in the intensive care unit for a couple of days.

Several months after his illness, Jesse's mother brought him into the office for a checkup. Amazingly, he showed no signs of any long-term complications. His hearing and eyesight were absolutely normal. He toddled around my office, babbling away, joyfully unaware of that night when his life teetered so very close to the edge. I looked at his parents and we all smiled.

UPDATE

According to the World Health Organization (WHO), Streptococcus pneumoniae *kills almost one million children less than five years of age worldwide every year; most cases are in developing countries. Pneumococcal (streptococcal) meningitis can occur in individuals of any age, but it is most common in very young children, as this encounter documents, and in middle-aged and elderly adults. Each year in the United States, the Centers for Disease Control and Prevention (CDC) report 175,000 hospitalized cases of pneumococcal pneumonia that often result from a common complication of influenza or measles. In addition, there are more than 50,000 cases of invasive disease, of which 3,000 to 6,000 cases each year are meningitis. Infected patients must be treated with antibiotics as soon as possible. The recent emergence of penicillin resistance in strains of pneumococci from many countries is very worrying and* will complicate antibiotic treatment. Pneumococcal meningitis represents one of the most common causes of death for which there is a vaccine. A pneumococcal conjugate vaccine, called Prevnar®, has been licensed since 2000 and is recommended for all children less than 24 months old and for children between 24 and 59 months old who are at high risk of disease. Older children and adults with risk factors are given the pneumococcal polysaccharide vaccine (Pneumovax® and Pnu-Immune®), which has been in use since 1977. The polysaccharide vaccines are a 23-valent vaccine, meaning it protects against 23 types of S. pneumoniae. The polysaccharide vaccine is recommended for use in all adults older than 65 years of age.*

QUESTIONS TO CONSIDER

1. Why did the physician believe that the disease affected the sixth pair of cranial nerves?
2. What are the classic signs of meningitis?
3. Why did the physician decide on a clinical diagnosis rather than a laboratory-based diagnosis?
4. What lead the physician to suspect the infection was bacterial rather than viral?
5. Why is the identification of white blood cells in the cerebrospinal fluid indicative of meningitis?

The pattern of infection and disease in the United States is changing, reflecting the upheavals in immigration patterns; more immigrants are arriving from regions of the world where diseases that are not common in the U.S. are prevalent. In addition, some of these diseases bring with them multidrug resistance, which makes treatment and cure an even more difficult objective. Indeed, the emergence of drug-resistant microorganisms threatens to make the disease in this encounter once again incurable.

THE SLEEPING GIANT

Tony Dajer

"No eating," the young woman said, pointing to her friend on the hospital bed. "For a week."

"British?" I asked.

"Russian," she replied, trilling the r like a commissar.

The patient was lying down, facing the wall. "Nausea and vomiting the last two weeks," read the nurse's note. "Also headache and fever. Is feeling better today." Her temperature was normal.

"Hello, Olga," I said, gently shaking her shoulder. "May I examine you?"

No response. Her friend whispered in her ear. Olga slowly sat up, eyes closed, black fur coat pulled tight. She was 38, with high cheekbones and blond hair. Although her chest sounds were normal, her belly seemed tender in the right upper quadrant, over the gallbladder. I tried to flex her neck.

"Ow," she cried.

Her headache and fever couldn't tell me much; they're common in many diseases. Throw in neck pain, however, and you must consider meningitis. But meningitis for two weeks?

Reprinted with permission from the May 2001 issue of Discover *magazine. Copyright © Discover Magazine. All rights reserved. For more information about reprints from* Discover, *contact PARS International Corp. at 212-221-9595.*

With no fever now? Maybe a brain tumor. Or gallbladder disease—a common cause of recurrent vomiting.

But results from a head CAT scan and gallbladder ultrasound proved negative.

I checked Olga's neck again.

"Ow," she said, exasperated.

"Please tell her," I said to the friend, "that we need to perform a spinal tap."

The mere suspicion of meningitis— an infection of the fluid lining of the brain and spinal cord— demands immediate diagnosis and intravenous antibiotics within minutes. A relatively benign viral infection of the meninges, the membranes that encase the spinal cord and brain, seemed the best bet; virulent bacterial meningitis does not putter for two weeks. And Olga didn't seem that sick. But a spinal tap was the only way to be sure.

I slipped the long spinal needle between the vertebrae of her lower back, then removed the thin stylet. Crystal-clear cerebrospinal fluid dripped into the sample tubes.

"It's probably fine," I told Terry, the nurse who assisted me. "Let's give her a gram of Rocephin"— a dose of preemptive antibiotics.

But the microscopic analysis said she was not fine: Olga's cerebrospinal fluid contained 176 white cells per cubic milliliter instead of the normal few. Most were mononuclear cells of the class that usually fights viral, not bacterial, infection. Ominously, the protein in the fluid was three times normal, and the glucose only half. That increase argued for bacterial meningitis. Yet the spinal tap yielded no sign of bacteria. Playing it safe, the admitting team gave Olga two more broad-spectrum antibiotics.

The next day Olga ate a hearty breakfast. But at noon, the ward nurse found her nearly comatose.

It made no sense. With massive antibiotics, Olga had been improving.

Her lethargy and confusion meant the infection might have spread past the meninges and infected neurons. Certain aggressive viruses, such as West Nile, can penetrate brain cells and wreak havoc. A more common culprit is herpes simplex virus, the cause of cold sores. Fortunately, herpes yields to an antiviral called acyclovir.

The residents performed another spinal tap. The count of mononuclear white cells had jumped to 900.

"Wow," I said. "TB?"

The tuberculosis bacillus, *Mycobacterium tuberculosis*, shrugs off antibiotics that kill other bacteria. And, unlike its cousins, it provokes attack by mononuclear white cells.

"We've sent the samples in for culture and testing," they said.

"Are you giving antibiotics?"

"No, infectious diseases says to wait."

The tuberculosis bacillus, once called "the captain of all these men of death," can be fiendishly elusive. It always enters through the lungs, but it doesn't always stay there. In about 15 percent of cases, the first signs of disease crop up elsewhere. And in half of those cases, chest X rays are normal. To get to the brain, the TB bacilli must evade the lungs' immune defenses and slip into the bloodstream. Once the bacteria reach the space beneath the outermost membranes of the brain, where the cerebrospinal fluid flows, they form clusters. On occasion, a cluster bursts, sending new bacilli through the brain's fluid-filled passages to establish new colonies. The process can smolder for months, causing inconstant, subtle symptoms doctors often miss.

The day after getting acyclovir, Olga seemed a little better. The results of her TB skin test, in the meantime, were inconclusive. The intern and resident, for their part, kept pointing to the low glucose in the cerebrospinal fluid, a finding more typical of tuberculosis than herpes. But the infectious diseases consult did not want to commit Olga to a six-month course of four potent antituberculosis drugs without more data.

So a PCR test, which amplifies a pathogen's DNA for identification, was on the way. It takes barely a week, compared with the six weeks it takes to culture TB bacteria. And PCR is extremely accurate. It also reliably detects herpes.

On the fifth day, Olga perked up. A third spinal tap showed only 120 white cells. But the glucose was still low, and the protein count high. The next day, Olga stopped speaking and needed to be restrained from falling out of bed. On day seven, she bounced back again. But another clue surfaced: Her husband said a cousin had TB meningitis four years earlier, just before Olga left for the United States.

And then we got confirmation: The PCR was positive for *Mycobacterium tuberculosis*. Olga immediately got four antibiotics that target TB. Two days later, she was sitting up, smiling. But she wasn't home free.

Over the next month, the nerve in her left eye that controls glancing sideways stopped working, making her see double whenever she looked left. Then she developed hydrocephalus— the accumulation of fluid in the brain— due to inflammation of the meninges. A combination of antibiotics and steroids kept that in check. Best of all, stroke, a dreaded complication, never struck.

A century ago, tuberculosis sickened one in five people. Today, more than 50 years after the advent of the first antibiotic against tuberculosis, the microbe still kills 2 million people around the world each year. In the United States, the prevalence of tuberculosis is low, but the global pandemic will not spare us. In 1993, foreign-born Americans accounted for 30 percent of tuberculosis cases in the United States; in 1998, the figure jumped to 41 percent.

Within a few weeks, Olga's left eye returned to normal. One month after her arrival, she was wheeled into the hospital lobby and left— wan and terribly thin— under her own steam.

"That type of infection used to be 100 percent fatal," the intern said as he watched her go. "And, boy, isn't that PCR slick?"

Sure, I mused, but an old-fashioned diagnostic guess can be even slicker.

UPDATE

There were 13,779 tuberculosis (TB) cases reported in the United States in 2006, representing a 2.1 percent decline from 2005 and the lowest recorded since national reporting began in 1953. California, New York, Texas, and Florida accounted for 48 percent of the national case total. Although the number of TB cases reported annually in the U.S. has decreased 48 percent since 1992, the annual case rate has slowed, from an annual average decline of 6.6 percent for 1993 to just over 3 percent in 2006. Foreign-born Americans again accounted for an increasing number of TB cases, from the 41 percent in 1998 to almost 60 percent in 2006. Tuberculosis deaths decreased by 1.7 percent to 646 deaths in 2005.

60   The Sleeping Giant

In 2006, the Centers for Disease Control and Prevention (CDC) for the first time included a case count of extensively drug-resistant TB (XDR TB) cases. Extensively drug-resistant TB is defined as having resistance to the first-line anti-TB drugs isoniazid and rifampin, with resistance to any fluoroquinolone and at least one of three injectable second-line anti-TB drugs (i.e., amikacin, kanamycin, or capreomycin). Three cases of XDR TB were reported in the U.S. in 2006.

QUESTIONS TO CONSIDER

1. How do the bacilli of this disease reach the meninges and the brain?
2. How did the polymerase chain reaction (PCR) help identify the microbe in this encounter?
3. What observation argued against Olga's illness being meningitis?
4. How does the tubercle bacillus enter the human body?
5. In what part of the United States population is the TB case count the highest? Why is this?

When diagnosing a disease, often the signs and symptoms are so clear that the diagnosis is a slam dunk. In other cases, however, the signs and symptoms may be very general or masquerade as another disease. This encounter is an example of such a mysterious disease, which if misdiagnosed, can be life threatening.

WHO'S THAT?

JOHN R. PETTINATO

Melissa's first brush with death, an automobile accident that put her in a coma, came when she was 15. After extensive rehabilitation, her youthful vitality returned. Her next bout with death came 10 years later, and this time I was afraid her luck had run out.

It was April, and the freshness of spring had just begun to wash away the gloom of winter. But this spring brought Melissa shaking chills, night sweats, and vomiting—the flu. Her prescription was simple: plenty of rest, plenty of fluids.

The plan worked until she became angry and mean-spirited, and lost interest in her personal appearance. "Just the flu," her doctor said to her anxious husband. Then she stopped eating, and the seizures she'd had since her car accident began to last longer. Her husband took her to the emergency room. He later told me that he was so sure that she would be admitted that he brought an overnight bag with toiletries and extra clothes. Again they were told it was just the flu, and Melissa was sent home.

Within days, she was almost catatonic, unable to do little more than drool. This time her husband, in an act of devotion

---

Reprinted with permission from the January 2002 issue of Discover *magazine. Copyright © Discover Magazine. All rights reserved. For more information about reprints from* Discover, *contact PARS International Corp. at 212-221-9595.*

that saved Melissa's life, refused to leave the emergency room until the attending physician agreed to admit her.

I was on call. I started by reviewing the lab data. The only hint of infection was her elevated white-blood-cell count. Where was the infection? Blood and urine cultures were negative, and her chest X ray didn't show any pneumonia. The clinical presentation, fever and mental-status changes, suggested a central nervous system disease. A lumbar puncture showed spinal fluid consistent with a viral infection. But another detail about the spinal-fluid analysis caught my eye—red blood cells, a sign of bleeding in the brain. Perhaps she had encephalitis. Many viruses can infect the brain, but I knew of only one with that signature: herpes simplex.

Herpes simplex viruses type 1 and type 2 tend to infect mucous membranes and the central nervous system. HSV-1 causes cold sores and 95 percent of all central nervous system infections. HSV-2 is associated with genital disease. HSV-1 spreads through contact with virus-laden saliva or sores. People usually get infected in childhood or adolescence. Sometimes there are no symptoms, but more often people have cold sores, or "fever blisters," in or around the mouth. After the first infection the virus lies latent in the trigeminal ganglia, a structure in the base of the brain that gives rise to the trigeminal nerve, which predominantly provides sensation to the face. For reasons that are not clear, the virus can reactivate years later and travel along the trigeminal nerve to the meninges, the coverings of the brain and spinal cord, at the base of the brain. From there it can launch an attack on its preferred target: the temporal lobes, brain regions just above each ear that help carry out the complex functions of hearing, learning, memory, and emotion.

Untreated herpes-based encephalitis can be fatal in up to 70 percent of cases. Malaise, fever, and headache herald its onset, often quickly followed by behavioral abnormalities, seizures, olfactory hallucinations, and bizarre or psychotic behavior—all symptoms of disease in the temporal lobes.

Fortunately, effective antiviral therapy is available, and early treatment reduces mortality to 30 percent. But making the diagnosis promptly is vital. And even with treatment, survivors of herpes encephalitis are almost never neurologically normal and will often experience amnesia, seizures, and anosmia, the loss of smell.

When I first examined Melissa, she was comatose. An electroencephalogram showed slowing of the normally brisk electrical activity of the brain. That finding fit with her comatose condition, but it is not typical of the waveforms often seen in herpes encephalitis. The only good news I had to share with her family was a normal CAT scan. I could test the cerebrospinal fluid for viral DNA using the polymerase chain reaction (PCR) technique to pin down the diagnosis of herpes simplex, but the results wouldn't be available for days.

Melissa was dying and something had to be done now. I followed my instinct and treated her for herpes encephalitis with acyclovir, an antiviral drug.

Two days later, the PCR results came back positive for HSV.

Melissa recovered, but her return home was like stepping onto the set of a movie she'd never seen. She couldn't find the guest bedroom, and she would often joke that her husband had moved it when she wasn't looking. She didn't remember that she had been taking tennis lessons. She didn't even recognize her tennis racket. She also had a bigger problem: She couldn't recognize faces. Pictures of high school friends were the faces of strangers. Patients with this condition, called prosopagnosia, can identify a face as a face, its parts, and even certain emotions, but they are unable to identify a particular face as belonging to a specific person. Prosopagnosics often do not recognize their own faces in the mirror, although they will recognize that they are looking at a face.

Still, Melissa didn't lose her knowledge of people's identities. She just couldn't count on using facial recognition to make identifications. Because humans are remarkably adaptive, patients like Melissa can often be taught how to compensate. Over time, Melissa learned to recognize people by context, such as where she last saw a person and what he or she was wearing. I saw this firsthand when I met her for a follow-up visit. Only when we were seated in the customary positions in my office did I see a glimmer of recognition flash across her face.

After several months of rehabilitation, Melissa was able to return to teaching. She leads a near-normal life. During my last visit with her, I said I was pleased at her recovery. With a twinkle in her eye, and as if to make light of it, she smiled slyly and said, "Dr. Pettinato, it's just a matter of recognition."

UPDATE

Herpes encephalitis is the most common cause of sporadic viral encephalitis. The herpes simplex virus type 1 (HSV-1), which accounts for 95 percent of all fatal cases of sporadic encephalitis, has a preference for the temporal lobes and, as described in this encounter, a range of clinical presentations, including meningitis, fever, and a severe, rapidly progressing form involving altered consciousness. Although adult herpes encephalitis accounts for 10 to 20 percent of viral encephalitis in the United States, the disease is rare; about 2 cases per million individuals are reported each year. The mortality rate in untreated patients is 70 percent. Among treated patients, the mortality rate is 19 percent, although more than 50 percent of survivors are left with moderate or severe neurological deficits. Besides the encounter that Melissa had, in his book Musicophilia, *Oliver Sacks relates the case of Clive Wearing, a distinguished English musician, conductor, and musicologist. In 1985, Mr. Wearing developed herpes encephalitis. Although he recovered, the virus so severely destroyed his memory that Clive could not recount anything about his life of 45 years. Yet, incredibly, he could perform, sing, and conduct a choir and a whole orchestra perfectly, with all the talent, sensitivity, and musical intelligence he had before the infection. However, once the performance was over, Clive would be confused and would have no idea that he had expertly conducted an orchestra.*

QUESTIONS TO CONSIDER

1. What effective viral therapy is available for this brain disease?

2. Propose some ways that Melissa had come in contact with the infectious agent?

3. What are the clinical signs of this mysterious disease?

4. What was the key sign that pinpointed the mysterious disease?

5. What infections and diseases are caused by herpes simplex virus type 1 (HSV-1) and type 2 (HSV-2)?

When one thinks of waterborne or foodborne outbreaks, one envisions drinking contaminated water from an unsanitary source or eating contaminated food at a restaurant. Sometimes, however, an infectious disease outbreak can originate in places one would not suspect, as this encounter illustrates.

JUST AN UPSET STOMACH?

CLAIRE PANOSIAN DUNAVAN

Not long after landing, an attendant on Flight 386 knew something was wrong. Between gurgles, pings, and cramps, her gut felt like a punching bag. She quickly disappeared into the lavatory.

Meanwhile, in a different part of town, a passenger from the same flight was also in distress. Her bowels were running like water. Suddenly it dawned on her: This intestinal bug was out of control.

In a third house in the sprawling metropolis, a 70-year-old man who had just flown 6,000 miles to visit his family checked his shaving kit to see if his diarrhea pills were still there. Well, maybe he should take a couple: This illness was worse than usual.

Three days later, the flight attendant and the sightseer were recovering. The grandfather was dead.

In January 1991, epidemic cholera surfaced in South America for the first time in the 20th century. Originating in coastal Peru and rippling inland, the toxic tide reached 14 Latin American countries and spread as far north as Mexico, infecting nearly half a million residents and killing 4,000.

But on Valentine's Day 1992, cholera was the last thing on the minds of the 356 passengers and crew on Aerolineas

Reprinted with permission from the July 2003 issue of Discover *magazine. Copyright © Discover Magazine. All rights reserved. For more information about reprints from* Discover, *contact PARS International Corp. at 212-221-9595.*

66 Just an Upset Stomach?

Argentinas Flight 386. The flight had been uneventful, departing Buenos Aires, Argentina, stopping in Lima, Peru, and then landing in Los Angeles. But 24 hours later, diarrhea struck six of the passengers. On February 16, 25 more were affected. By February 17, the toll was up to 54 infected and one dead.

Sometimes I wonder if those early sufferers realized they were part of an outbreak. They certainly had no way to compare notes, dispersed halfway around the globe. On February 19, however, five Los Angeles-area hospitals reported stool cultures growing the bacterium *Vibrio cholerae*—all from travelers on Flight 386. That's when the warning bell sounded, and county health officials knew they had to try to find the remaining 351 passengers and crew. From a medical standpoint, they were late. Although most cholera patients suffer only mild to moderate diarrhea, others have been known to drop dead in a day.

There was a second reason to track the entire planeload of people. In order to pinpoint the source of the outbreak—presumably cholera-laced food served on board—health authorities had to survey as many passengers as possible, both with and without symptoms. At the same time, another question gnawed: Could public-health detectives find the tainted food before other flights were exposed?

In Los Angeles, the crisis couldn't have come at a worse time. On the same weekend the cases broke, a fire closed the downtown offices of the county's acute communicable disease unit. In makeshift quarters miles from home, the front-line squad launched an investigation without computers, files, or phones. Looking back, the unit's chief, Laurene Mascola, recalls one silver lining. "The good thing . . . was that all of us were in one big, open area. And so we worked very efficiently, as we had few other distractions."

They still faced a tall order: reviewing customs declaration forms to identify heads of household on the flight, calling and faxing to determine who else was on board, talking to crew members (this required delicate negotiations with the airline), and cajoling caterers in Buenos Aires and Lima for menus, all the while juggling daily press conferences with news-hungry reporters.

Somehow, the wobbly effort succeeded. When investigators finally reached the sick and the well, compared their food diaries, and tested their lab samples, a shrimp salad prepared in Lima was the smoking gun. The other good news? Only Flight 386 had served the microbe-laden dish. On the other hand, of 194 passengers and crew who submitted blood or stool samples, 100 had undeniable evidence of recent cholera infection.

As a tropical-medicine specialist, I first heard about the rash of cholera from newspapers. It was a big story because the circumstances of the outbreak were so unusual. Nobody expects to get cholera on an airplane.

The disease typically occurs in regions where diarrheal illnesses spread easily because of inadequate sanitation. As many as 2 million infants and toddlers in developing countries still succumb to diarrhea every year. Most of these deaths are not from cholera, however. While the intestinal pathogens *Rotavirus* and toxin-producing *Escherichia coli* are pervasive in hygiene-poor countries, cholera is more often a sporadic wildfire that snakes through the high-risk settings and even crosses oceans. When it's not wreaking havoc in a favela or a refugee camp, the organism's survival strategy is to hide out in brackish waters affixed to the horny exoskeletons of plankton and shellfish, an ancient form of bacterial hibernation.

What cholera and harmful *E. coli* have in common is a poison (also known as an enterotoxin) that binds to the inner lining of the small intestine. As a result, fluid and electrolytes are secreted rather than absorbed, and the affected gut gushes like a broken fire hydrant. Cholera toxin in particular is so potent that some victims purge as much as a liter per hour of nearly clear diarrhea, known to medical officers during the British raj as rice-water stool. Even today, if patients in this subgroup don't receive intravenous fluids or down massive amounts of a glucose-electrolyte solution like that found in popular sports drinks, they can die from desiccation.

Sadly, the grandfather from Flight 386 turned out to be a textbook example of a worst case. Three days after becoming infected, his blood pressure became dangerously low, his pulse weak and thready, his bowel bulging with fluid, while paradoxically, his oral membranes were so dry that one examiner's tongue depressor stuck to the roof of his mouth. By the time he reached an intensive care unit, he was comatose, and his tissues were dangerously acidic. Despite heroic rescue efforts, he died within hours.

68 Just an Upset Stomach?

Several years later, I received a phone call from the director of a nearby family-practice training program. "How would you like to discuss a case of traveler's diarrhea in next week's Grand Rounds?" he asked.

"Sure," I said. "Just fax me the write-up."

Soon I was reading about a middle-aged woman who had recently returned from El Salvador. On her flight to Los Angeles, she developed abdominal pain, vomiting, and profuse watery diarrhea. After three days, too weak to even hold up her head, she was brought to the hospital. Her lab values told the rest of the story. Her blood was highly concentrated, her potassium dangerously low, and her kidneys had nearly shut down.

This has got to be cholera, I thought. I wonder how the residents handled the case?

The following week, my question was answered when two young doctors-in-training stepped to the podium. What they described next would make any residency director glow with pride. Suspecting cholera from the outset, they infused six liters of an intravenous glucose-electrolyte solution known as Ringer's lactate solution, inserted a urinary catheter and a rectal tube, and then meticulously charted I's and O's (medicalese for "intake" and "output") through all orifices in order to replace the exact amount of fluid needed. They also started doxycycline, an antibiotic that lessens cholera's overwhelming fluid loss as well as its bacterial stool count. Their management was flawless. Slowly, the patient's kidneys began to work again, and her diarrhea abated. Five days later, the patient walked out of the hospital vibrant and well.

Of course, a little credit also goes to Flight 386. Where I live, it made cholera a household word.

UPDATE

Cholera has been very rare in the developed nations of the world for the last 100 years. However, a few cases occur every year in the United States in persons who travel to a malarious part of the world or eat contaminated food brought back by travelers, as seen in this encounter. In 2005, the largest number of laboratory-confirmed cases of Vibrio cholerae *infection were reported since 1998. There were nine cases and none of the patients hospitalized for cholera died. In 2005, about 36 percent of cases were acquired outside the*

United States; foreign travel and consumption of undercooked seafood remains the major sources of illness. Another 36 percent were attributable to consumption of domestic seafood, specifically from contaminated crabs harvested from the U.S. Gulf Coast after Hurricane Katrina. In fact, crabs harvested from the U.S. Gulf Coast are a common source of cholera, especially during warmer months, when environmental conditions favor the growth and survival of V. cholerae *in brackish and coastal waters. For the other 28 percent, no source was identified. On a global scale, cholera outbreaks frequently occur as people migrate into urban centers of developing countries. Such migrations strain existing water and sanitation infrastructure, which increases disease risk. Almost 132,000 cases, including more than 2,200 deaths in 52 countries, were reported by the World Health Organization (WHO) in 2005. In addition, man-made and natural disasters can increase the risk of cholera epidemics and produce explosive outbreaks with high case-fatality rates. For example, following the Rwanda genocide in 1994, outbreaks of cholera in crowded refugee camps in Goma, the Congo, caused at least 48,000 cases and almost 24,000 deaths within one month.*

QUESTIONS TO CONSIDER

1. What solution should be given to someone suffering from cholera?
2. When trying to track the source of the outbreak on Flight 386, why were passengers who had no symptoms interviewed?
3. Where is cholera typically found?
4. What was the source of the airborne outbreak?
5. Explain how *Vibrio cholerae* causes cholera.

Foodborne illnesses and outbreaks caused by toxin-producing bacteria are increasing in the United States. Since some of these intoxications or infections may be life-threatening, emergency room physicians must be ready to quickly diagnose the illness and provide the patient with immediate therapy and medication. If the patient is an infant, as in this encounter, a rapid and correct diagnosis is of the essence.

BAD BLOOD

MARK COHEN

Cheryl Taylor, the attending physician at the emergency room, looked up from a chart she was studying. She had called me in to evaluate one of my patients, Casey Morita, a 2-month-old who had arrived looking very dehydrated.

"Hi, Mark, thanks for coming in," she said. "As I told you on the phone, I've given her a total of 40 cc's per kilogram of IV fluid. She's looking a little perkier, but her chemistries are really out of whack."

When I last saw Casey, just one week ago, she was perfectly healthy. So I was puzzled and worried as I went in to examine her.

Casey was sleeping in her mother's arms in the curtained-off cubicle. A brief glance and a touch told me that Cheryl had done a good job of rehydrating her: Casey's color was good, and her skin did not have the dry, wrinkly feel of severe dehydration. Casey's mother said her daughter had been well until two days ago, when she began having frequent, watery bowel movements. Last night Casey couldn't keep down any of her formula or the commercial electrolyte solution her mother had

---

Reprinted with permission from the December 2002 issue of Discover *magazine. Copyright © Discover Magazine. All rights reserved. For more information about reprints from* Discover, *contact PARS International Corp. at 212-221-9595.*

offered her. She had no fever, no congestion or cough, no rash, and no one else at home had been sick. It sounded like a simple gastrointestinal infection. But why would she get so sick so fast?

A careful examination turned up nothing unusual. To figure out what was going on with her, I'd have to look at the levels of ions and molecules in her blood.

I first looked at the electrolytes: sodium, potassium, chloride, and bicarbonate. These four ions keep all the myriad cellular processes working properly, and bicarbonate in particular maintains the acid-base balance of the blood. The levels of these four ions in the blood are usually kept in close balance by the kidneys and endocrine system. Values outside the normal range could indicate excessive fluid loss, or a serious breakdown in the control system. The lab results said something was seriously wrong. Her sodium was 126 milliequivalents per liter (normal range 136-145), potassium 6.9 (3.5-5.5), chloride 87 (94-110), and bicarbonate 12 (19-28). The low bicarbonate told me her blood was much too acidic. If her potassium went much higher, it could cause her heart to beat irregularly or even stop altogether.

Diarrhea and dehydration can wreak havoc with these ions, but two other blood test results were even more worrisome: the BUN (blood urea nitrogen) and creatinine. Urea and creatinine are cellular waste products that are normally removed by the kidneys. Casey's BUN was 52 milligrams per deciliter; the normal range is 5-15. Her creatinine was 3.7; normal is 0.1-0.6. These values, plus the abnormal electrolytes, suggested the kidneys were not doing their job. The wastes were building up.

I asked the nurse to check her blood pressure. It was 135/82—seriously elevated for a 2-month-old. That, too, meant her kidneys were not working. The kidneys secrete renin, a hormone that helps maintain normal blood pressure. When lack of local blood flow interrupts kidney function, the kidneys release lots of renin in a misguided effort to boost blood pressure and get more blood flowing.

I called the children's hospital and asked them to send an ambulance and a team from the intensive care unit for Casey. This little girl, who had been perfectly healthy just a few days before, was in danger of dying from kidney failure—and I didn't know why.

One diagnosis did come to mind, but it was for a condition I had only read about and never seen. I checked the results of some of her other tests, and they tended to confirm my suspicion. First, Casey was anemic; the level of hemoglobin in her blood was low. So was the number of red blood cells. It's normal for a 2-month-old to have a mild degree of anemia, because she is still building her blood supply. But Casey's red blood cell count was even lower than expected. And the number of platelets, the tiny cellular fragments that clump up to form blood clots, was also low. That's not normal at any age.

Anemia, low platelets, and kidney failure, together with a history of diarrhea, suggested that Casey might have hemolytic uremic syndrome. This condition is usually caused by an infection, and the cascade of effects can be catastrophic. When a certain strain of a common bacterium infects the digestive tract, it produces a toxin—verotoxin—that passes from the intestine into the bloodstream. The toxin damages the endothelial cells, which make up the inner lining of blood vessels. The cells become swollen, narrowing the diameter of the vessels. Then, as platelets glom onto the damaged endothelial cells, the platelet count goes down, and the vessels narrow even more. In addition, the red blood cells themselves are damaged as they try to squeeze through narrowed capillaries; the result is a hemolytic (literally, "broken blood") anemia.

A dense network of capillaries supplies blood to the kidneys, so they are particularly vulnerable to the toxin's effects. The swelling of the capillaries and the accumulation of damaged red blood cells and platelets can become so severe that blood simply cannot flow through the kidneys. The production of urine slows to a trickle or stops completely. Waste products, including urea, build up in the bloodstream, causing a condition known as uremia—urea in the blood. It's the combination of hemolytic anemia (indicating red blood cell breakdown) and uremia (indicating kidney failure) that gives the condition its name.

First identified in the 1980s, the syndrome made headlines in 1992, when 453 people in Washington State fell ill. Forty-five of them developed hemolytic uremic syndrome, and three died. Many of those who got sick had visited a fast-food chain and eaten hamburgers. The illness was probably spread through undercooked beef contaminated with *Escherichia coli*

O157:H7, a bacterium that produces verotoxin. After several more outbreaks, we learned that thorough cooking of meats and careful attention to safe food-handling practices can reduce or eliminate the danger from these bacteria. We also learned that other bacteria can cause the syndrome. Casey did have *E. coli* O157:H7 in her stool, and this almost certainly caused her diarrhea and subsequent illness. But we don't know how she got it, because her only food was infant formula, which did not contain the bacterium. Ongoing investigation by the state health department may give us an answer.

By the time Casey reached the intensive care unit, she was in complete renal failure. Her doctors had to take over the job of the kidneys, using peritoneal dialysis. A tube inserted into her abdomen pumped fluid in and out several times a day, allowing the waste materials to be excreted across the large peritoneal membrane in the abdomen instead of through the kidneys. The capillary swelling gradually improved. Within a few days, Casey's body began to produce urine, and soon she no longer required dialysis.

Several weeks have passed, and Casey's kidneys have still not fully recovered. She is on a special formula and needs medication to control her blood pressure and chemical imbalances. While she seems to be growing well, we don't know whether her kidneys will return to normal or how long that might take. Some people with this syndrome have regained normal kidney function only to have their kidneys deteriorate years later. Casey survived. Now we hope she can avoid lasting scars.

UPDATE

The classic childhood case of hemolytic uremic syndrome (HUS), as described in this encounter, occurs in about 2 to 7 percent of all Escherichia coli O157:H7 *infections. In fact, it is one of the most common causes of sudden, short-term kidney failure in children. Most cases of HUS occur after an infection of the digestive system by* E. coli O157:H7 *bacterium, which is found in foods like meat, dairy products, and juice that are contaminated. In 2005, there were 221 HUS cases reported by the Centers for Disease Control and Prevention (CDC). In 2006, an epidemic of harmful* E. coli O157:H7 *emerged in the United States due to contaminated spinach. Most affected individuals were adults and 171 known cases* were reported to the CDC, including 27 cases of HUS; one adult in Wisconsin died. Then in 2007, more than 21 million pounds of ground beef were recalled after it was linked to an E. coli outbreak that sickened at least 40 people in the U.S. The CDC has stated that E. coli is responsible for sickening more than 73,000 people every year in the U.S. About 60 of the stricken individuals will die from HUS. In fact, recalls of E. coli–tainted meat have doubled in 2007, and outbreaks of the foodborne illness are on the rise as well.

QUESTIONS TO CONSIDER

1. Why did the initial diagnosis suggest a gastrointestinal infection?

2. What were the indicators that suggested a diagnosis of hemolytic uremic syndrome?

3. What is the role of bicarbonate ions in the blood, and what does a low level of the ions indicate?

4. What blood tests, besides the electrolytes, suggested that Casey's kidneys were not functioning properly?

5. What type of chemical agent caused Casey's illness, and how does the agent damage blood vessels?

Many of us see weekend yard work as a task (onerous and pleasurable) that must be done to keep the yard looking nice and prevent it from becoming an eyesore to the neighbors. The trimming and cutting often can be exhausting work and sometimes dangerous if one is climbing on ladders or using electrical equipment. We don't usually equate yard work with contracting a potentially lethal infection, yet such encounters, though rare, do occur, as described below.

A TASK IN THE YARD TURNS LETHAL

CLAIRE PANOSIAN DUNAVAN

My husband and I live in a cottage in the foothills of Los Angeles, where nature feels very close. Beyond our front door, an ancient flowering vine overhangs a brick porch. Tangled up within the vine is a whole world in miniature: abandoned birds' nests, dangling spiderwebs, powdery organic deposits. For years, we marveled at the vine's ecosphere—but we never grasped its intense biological power until one spring morning I will never forget.

"I'm going outside," my husband said. "I feel like pruning."

After finding his clippers, Patrick started to yank and trim the tangled greenery. Then I heard a loud, strangled cough. "Yech!" he exclaimed, violently stomping and shaking himself. "I feel like I just inhaled toxic waste—my lungs are on fire!"

Because Patrick has asthma, sudden fits of wheezing and shortness of breath are nothing new to him. This was different. Some dusty emanation from the vine had triggered a

Reprinted with permission from the August 2007 issue of Discover *magazine. Copyright © Discover Magazine. All rights reserved. For more information about reprints from* Discover, *contact PARS International Corp. at 212-221-9595.*

fierce pain from his trachea to the deepest cul-de-sac of his lungs. An hour after his noxious gulp of air, though, he felt better. I figured the worst was over.

Another day passed, and my husband's nose began to run. He was also clearing his throat more than usual. He was coming down with a cold, we decided. My main concern was its timing: Later that week, we were supposed to leave for New York.

Before our departure, Patrick armed himself with antihistamines and an inhaler. He felt OK on the flight, but the next day he was tired after walking just a few blocks. We chalked it up to the bustle of Manhattan and rushing to make the curtain of a play.

Finally, while we were seated in our hotel room later that night, alarm bells went off. My husband's face was flushed, his pulse was fast, and he said he felt as if he was breathing through "a barrel of phlegm." Could he have pneumonia? How had I overlooked such an obvious possibility?

"I'd better listen to you," I said quietly.

I pressed the side of my head against Patrick's back to listen as he breathed in and out. The racket on both sides—like the groan of a low-pitched, badly played accordion—was unlike any sound I remembered from previous asthma attacks. Through the cartilage of my ear, I could practically feel the rattle of secretions in his airways. Think again, my doctor brain commanded. Whatever was making my husband sick, it was no ordinary wheezing, cold, or even pneumonia.

We pondered our next move. Should we hunt down a doctor? Weighing the pros and cons of visiting a strange local emergency room versus limping back home to Los Angeles, we opted for the latter. Meanwhile, Patrick doubled his standard asthma doses while I lined up an urgent medical appointment. En route to JFK Airport, I almost asked the taxi driver to turn around. But Patrick shook his head.

At last, we sat face to face with Roy Young, our internist in Los Angeles. A seasoned pro, he quickly listened to the din in Patrick's chest. Then he pulled out his prescription pad.

"We'll get a chest X-ray and blood work, of course, but you're starting on steroids right now," Roy said. "And azithromycin."

"What about this stuff I'm coughing up?" By then, Patrick was producing tablespoonfuls of thick sputum.

"Let's do a culture," Roy replied. "It might show something interesting."

Two days later, the agar plates streaked with Patrick's sputum began to sport not some nasty strain of lung-loving bacteria but rough patches of gray-green mold. Viewed through a microscope, the slim branching stalks topped by swollen vesicles and spores answered our burning question. My husband's airways had become a hothouse for a fungus named *Aspergillus fumigatus*.

Allergic bronchopulmonary aspergillosis—ABPA for short—is extremely rare. While most people can inhale the fungus without the slightest discomfort, one subgroup is particularly vulnerable: chronic asthmatics. Something about the mucus in their lungs fuels the growth of the hardy, ubiquitous fungus for which the syndrome is named.

Once *Aspergillus* starts to thrive in the bronchial tree of an asthmatic, it spurs the production of even more mucus. This, in turn, worsens the victim's airway spasms. If the fungus is not quashed by powerful anti-inflammatory drugs, this vicious cycle can go on for months. Even then, in some sufferers, the inflammation triggered by the fungus continues to damage the normal spongy lacework of the lungs.

You could say that isolating *Aspergillus* from Patrick's sputum within days of his falling ill was a great stroke of luck; after all, most ABPA patients wait much longer before they are diagnosed. But to me, it was also sobering. As an infectious diseases doctor, I knew *Aspergillus's* nasty handiwork. In immunocompromised patients (leukemics or organ transplant recipients, for example), the fungus can eat into lung tissue and blood vessels. The resulting hemorrhagic pneumonia—bleeding in the lungs—is often fatal. In some patients, the lungs become scarred or riddled with "fungus balls"—tangles of fungal filaments the size of golf balls. Major thoracic surgery may be required to remove them.

ABPA is far less threatening, but it still requires patience and expertise. As soon as he suspected the diagnosis, our internist referred Patrick to a pulmonary specialist, Mike Roth, who then ordered a CAT scan and a blood test of his immunoglobulin E (IgE) levels to determine the intensity of the immune response. The CAT scan showed the thickened, distorted airways and cloudy patches that are hallmarks of the syndrome. The serum IgE—a direct measure of allergy-induced antibodies—was also up, confirming a pattern typical of immune activation.

Now with all the evidence he needed before him, Mike calmly explained that a full-bore attack on the ABPA inflammation and its root cause offered the best chance for complete recovery. He extended the course of steroids to three months, warning Patrick about side effects like jangled nerves, insomnia, and fluid retention. In addition, he prescribed a month's worth of an antifungal drug that produced an even more exotic reaction: It gave Patrick's vision a temporary blue-green sheen.

Week by week, Patrick's airways slowly cleared. The final proof came later that year when he returned to spending long, hard days on his feet as a film director. Today his energy is high, his asthma is back to baseline, and his chest is quiet once more.

As for grooming the vine, that is permanently off his list of chores. Patrick is also cautious about other risky pastimes—like mowing the lawn, chopping wood, or hanging out in a damp basement—that might reexpose him to *Aspergillus* spores. Once an ABPA victim, always an ABPA victim, some experts say. In other words, rather than inducing protective resistance, repeat encounters with the fungus often trigger the same illness all over again.

When I recently took a whack at vine pruning, I could feel my chest begin to burn. Thank goodness I don't have asthma, I thought, vowing to bring home a mask to wear the next time I got the urge.

UPDATE

Aspergillus fumigatus, *the cause of allergic bronchopulmonary aspergillosis (ABPA), is ubiquitous in the environment. The fungus has no known sexual cycle and reproduces by producing asexual conidiospores that are released into the environment. A. fumigatus can, therefore, be found in soil and in decomposing plant litter as this encounter describes. No national surveillance for ABPA or other aspergilloses exists. Being very uncommon, ABPA is listed as a "rare disease" by the Office of Rare Diseases of the National Institutes of Health. This means that ABPA affects less than*

200,000 Americans. As in this encounter, certain people who inhale the fungal spores may become sensitized and develop a chronic allergic reaction. The spores will germinate and the fungus will grow at human body temperature. Usually, the fungus is quickly eliminated by the immune system of healthy individuals. If an infection takes hold, however, the illness differs from typical pneumonias caused by bacteria, viruses, and most fungi, in that A. fumigatus does not actually invade or destroy the lung tissue. Instead, the fungus colonizes the mucus (sputum) in the airways of people who produce increased amounts of mucus (e.g., people asthma or cystic fibrosis). Colonization causes the alveoli, the tiny air sacs of the lungs, to become packed primarily with a type of white blood cell called eosinophils and with increased numbers of mucus-producing cells. Treatment typically involves administration of the corticosteroid prednisone and the antifungal drug itraconazole. As the disease is controlled, the eosinophil and antibody levels usually fall, but they may rise again as an early sign of flare-ups, so periodic monitoring of the patient may be needed.

QUESTIONS TO CONSIDER

1. What indications suggested that the "dusty emanation" Patrick inhaled had triggered more than simply an asthma attack?
2. What can result in an immunocompromised patient who has inhaled the fungus?
3. What was seen in the agar plates streaked with Patrick's phlegm?
4. What are "fungus balls?"
5. What does the presence of serum IgE antibodies indicate?

Glossary

Abscess A pus-filled cavity resulting from inflammation and usually caused by bacterial infection.

Acyclovir An antiviral drug that blocks DNA replication and is used to treat herpes simplex virus infections.

Aerobic Referring to the presence of oxygen gas; one form of metabolism requiring oxygen for activity.

AIDS (acquired immunodeficiency syndrome) A disease of the immune system, caused by infection with the retrovirus HIV, which destroys certain immune cells and is transmitted through blood or bodily secretions such as semen.

Allergic bronchopulmonary aspergillosis (ABPA) A hypersensitivity response to the fungus *Aspergillus fumigatus*, which causes a spectrum of lung diseases commonly called aspergilloses.

Alveolus A tiny thin-walled air sac found in large numbers in each lung, through which oxygen enters and carbon dioxide leaves the blood.

Anaerobic Referring to the absence of oxygen gas; one form of metabolism not requiring oxygen for activity.

Anatomical diagnosis The identification of an illness or disorder in a patient by locating the physical site of symptoms.

Anemia A blood condition in which there are too few red blood cells or the red blood cells are deficient in hemoglobin, resulting in poor health.

Anosmia The absence or loss of the sense of smell.

Antibiotic A chemical substance, derived from bacteria or fungi, that can kill or inhibit the growth of bacteria.

Antibody A protein produced by certain immune cells in the body in response to the presence of a foreign substance.

Antimony A toxic crystalline element that occurs in metallic and nonmetallic forms.

Antiserum A blood fluid without clotting agents and containing one or more ready-made antibodies that can provide immunity against a disease.

Asthma A long lasting illness involving the respiratory system where the airways occasionally constrict, become inflamed, and are lined with excessive amounts of mucus, often in response to one or more allergic triggers.

Bacillary angiomatosis A bacterial infection characterized by the proliferation of blood vessels, resulting in the formation of tumor-like masses in the skin and other organs.

B cell A type of white blood cell that is formed in bone marrow and is involved in the production of antibodies.

Bell's palsy A type of facial paralysis.

Benign Referring to a non-cancerous (non-spreading) tumor, or a mild condition of an illness.

Bile The yellowish brown or green fluid produced in the liver, stored in the gallbladder, and secreted into the small intestine, where it aids in emulsifying fats.

Biliary atresia Malformed bile ducts in the liver.

Bilirubin A red bile pigment that is an intermediate product in the breakdown of hemoglobin in the liver.

Brackish Referring to a mixture of fresh and salt water.

BUN (blood urea nitrogen) A measure of the amount of nitrogen in the blood that comes from urea.

Burkitt's lymphoma A malignant tumor of the lymphatic system, specifically in B cells.

Catatonic Referring to an emotional disturbance characterized by rigidity of the muscles.

Caustic A substance historically thought to corrode or burn away disease.

Cerebellum The portion of the brain consisting of two hemispheres connected by a thin central region and serving to control and coordinate muscular activity and maintain balance.

Chronic Used to describes an illness or disease that lasts over a long period.

Clotting factor A substance in the blood that is essential for blood to coagulate.

Complement protein One of a set of proteins in the bloodstream that, together with antibodies, recognize and attack foreign cells such as bacteria.

Compromised immunity Referring to an individual whose immune system cannot mount a normal fight against infectious disease.

Conidiospore An asexual, non-motile, haploid spore of some fungi that can develop into a new organism if conditions are favorable.

Contagious Referring to a disease that can be transmitted from one person to another either by direct contact with the person or by indirect contact, such as by contact with clothes.

Colitis Inflammation of the colon.

Creatinine A breakdown product that is mainly filtered by the kidney and thus a measure of renal function.

CSF (cerebrospinal fluid) The colorless fluid in and around the brain and spinal cord that absorbs shocks and maintains uniform pressure.

CT scan A radiological scan in which cross-sectional images within a part of the body are formed using computerized techniques and are shown on a computer screen.

Cupping A historical medical practice in which a cupping glass created a vacuum and thus would increase the blood supply to an area of the skin.

Decision tree A flow diagram or a logical sequence of steps for solving a problem (algorithm) used in making diagnostic decisions and for treating a patient.

Dexamethasone A synthetic steroid used to treat inflammatory conditions.

Diagnosis The identification of an illness or disorder in a patient through an interview, physical examination, medical tests, and other procedures.

Differential diagnosis A systematic method used by clinicians to identify the disease causing a patient's symptoms.

DNA probe A fragment of DNA of variable length (usually 100 to 1000 bases long), which is used to detect the presence of nucleotide sequences (in a pathogen) that are complementary to the sequence in the probe.

Ectopic pregnancy The development of a fertilized egg outside the womb, such as in a fallopian tube.

Electroencephalogram A visual record of the electrical activity of the brain.

Electrolytes Ions [sodium (Na^+), potassium (K^+), calcium (Ca^{2+}), magnesium (Mg^{2+}), chloride (Cl^-), phosphate (PO_4^{3-}), and bicarbonate (HCO_3^-)] found in cells or the blood.

Encephalitis An inflammation of the brain, usually caused by a viral infection.

Endemic Referring to a disease that occurs in a specific geographic area.

Endothelial cell A cell that forms a layer lining the inside of some body cavities, such as the blood vessels.

Enteritis An inflammation most commonly of the small intestine.

Enterotoxin A bacterial poison associated with food poisoning that causes vomiting and diarrhea.

Eosinophil A type of white blood cell that plays a part in allergic reactions and the body's response to parasitic diseases.

Epidemic An disease that spreads more quickly and more extensively among a group of people than would normally be expected.

Etiological diagnosis The identification of an illness or disorder in a patient through the identification of the causes and origins of disease.

Exoskeleton The hard covering on the outside of such organisms as crustaceans and insects that provides support and protection.

Fallopian tube One of the two narrow tubes through which a human egg passes from one of the ovaries to the womb.

Fibrin An insoluble protein produced in the liver that aids in blood clotting.

Flagellate Referring to a microorganism with long thin cellular appendages called flagella that are used for movement.

Guillain-Barré syndrome A disease affecting the peripheral nervous system due to an immune response to foreign antigens.

Hemolytic uremic syndrome (HUS) A disease characterized by a loss of red blood cells through cell destruction, acute kidney failure, and a low platelet count.

Hemorrhagic Referring to the escape of blood through ruptured blood vessel walls.

HIV (human immunodeficiency virus) Either of two strains of a retrovirus, HIV-1 or HIV-2, that destroys specific groups of immune cells, the loss of which causes AIDS.

Host An organism in or on which another organism, especially a pathogen or parasite, lives.

Hydrocephalus An increase of cerebrospinal fluid around the brain, resulting in an enlargement of the head in infants.

Hydrophobia An extremely intense fear of drinking water, characteristic of rabies.

Hyoid bone A U-shaped bone positioned at the base of the tongue and that supports the tongue and its muscles.

Hypochondrium The upper part of the abdomen under the lowest ribs of the thorax.

Immunoglobulin E (IgE) The class of antibody that plays an important role in allergies (hypersensitivities).

Inflammation The swelling, redness, heat, and pain produced in an area of the body as a reaction to injury or infection.

Inflammatory response *See* **Inflammation.**

Intussusception A sliding of a portion of the bowel, creating swelling that leads to obstruction.

Itraconazole An antifungal and antiprotozoal drug.

Koplik's spots The formation of small specks in the mouth as a result of a measles inflection.

Larynx The cartilaginous portion of the respiratory tract behind the tongue and on top of the trachea.

Leukocyte A type of white blood cell.

Lumbar puncture *See* **Spinal tap**.

Lymphocyte A key type of white blood cell of the immune system that either produces antibodies or attacks infected cells.

Macrophage A type of white blood cell present in blood, lymph, and connective tissues that aids in the removal of harmful microorganisms and foreign material from the bloodstream.

Malignant Referring to a tumor that invades the tissue around it and may spread to other parts of the body.

Meninges The membranes surrounding and protecting the brain and the spinal cord.

Meningitis A serious, sometimes fatal illness in which a viral or bacterial infection inflames the meninges, causing symptoms such as severe headaches, vomiting, stiff neck, and high fever.

Meningococcemia The spreading of the meningococcal bacterium *Neisseria meningitidis* through the blood.

Meningococcus The bacterial species *Neisseria meningitidis* that is responsible for a form of meningitis.

Monocyte A type of white blood cell, formed in the bone marrow and in the spleen, that has a well-defined cell nucleus and consumes large foreign particles and cell debris.

Mononuclear cell *See* **Monocyte**.

Mononucleosis An infectious disease caused by a virus, producing fever, swelling of the lymph nodes, sore throat, and increased lymphocytes in the blood.

Nasopharyngeal carcinoma A cancer of the epithelial tissue in the nose and pharynx.

Neurasthenia A condition characterized by chronic mental and physical fatigue and depression; the "blahs."

Omentum A membranous fold that covers the intestines and connects them to the liver.

Opportunistic infection An illness occurring when a microorganism that is not normally serious becomes pathogenic or life-threatening when the host has a low level of immunity.

Palpable Referring to something that can be felt by the hands during a medical exam.

Pancolitis An inflammation throughout the colon.

Parasite An organism that lives on or in another, usually larger, host organism in a way that harms the host.

PCR (polymerase chain reaction) test A technique used to replicate a fragment of DNA and produce very large quantities of that sequence.

Pelvic inflammatory disease (PID) An inflammation of a woman's reproductive organs in the pelvic area, which can lead to infertility.

Peritoneal membrane *See* **Peritoneum**.

Peritoneum The membrane that lines the abdomen.

Plasma The fluid portion of blood and lymph that excludes the blood cells.

Platelet A very small colorless disk-shaped cell fragment in the blood that plays an important part in blood clotting.

Pneumonia An inflammation of one or both lungs, usually caused by infection from a bacterium, virus, or fungus.

Poultice A warm, moist preparation placed on an aching or inflamed part of the body to ease pain, improve circulation, or hasten the expression of pus.

Prednisone A synthetic steroid hormone used to treat allergies.

Pressor Referring to an increase in blood pressure.

Prognosis A medical opinion as to the likely course and outcome of a disease.

Prosopagnosia A disorder of face perception where the ability to recognize faces is impaired, while the ability to recognize other objects may be relatively intact.

Pseudomembranous colitis An infection of the colon often, but not always, caused by the bacterium *Clostridium difficile*.

Pus The yellowish or greenish fluid that forms at sites of infection, consisting of dead white blood cells, dead tissue, bacteria, and blood serum.

Renin A kidney enzyme that breaks down proteins and helps regulate blood pressure.

Reservoir An organism in which a pathogen or parasite lives and develops without damaging it, but from which the pathogen or parasite passes to another species that is damaged by the pathogen or parasite.

Ringer's lactate solution A solution of balanced blood salts and minerals that is intended for intravenous administration.

Sepsis A condition caused by the presence of bacteria, such as *Escherichia coli*, and their toxins in the bloodstream.

Septic Referring to or involving the presence of bacteria in the blood.

Septic shock A serious medical condition that occurs when an overwhelming infection leads to low blood pressure and low blood flow; leading to decreased oxygen delivery, it can cause multiple organ failure and death.

Shock *See* **Septic shock**.

Sign An indication of the presence of a disease or disorder, especially one observed by a doctor but not apparent to the patient; for example low grade fever, high blood pressure.

Specific treatment Refers to a clinician's specific treatment of the diagnosed disease and hopefully affecting the final outcome.

Spinal tap A surgical procedure that involves drawing spinal fluid using a hollow needle or tube.

Sputum A substance such as saliva, phlegm, or mucus coughed up from the respiratory tract and usually ejected by mouth.

Strabismus A condition in which the eyes are not properly aligned with each other.

Symptom An indication of some disease or disorder, especially one experienced by the patient; for example, pain, headache, or itching.

Symptomatic treatment Treating the symptoms of a ill patient.

Syndrome A group of signs and symptoms that, taken together, characterize a specific disease or disorder.

T cell *See* **T lymphocyte**

Temporal lobe Either of two lobes of the brain, located on the side of each cerebral hemisphere, that contain the auditory centers responsible for hearing.

T lymphocyte A type of white blood cell that matures in the thymus and is essential for combating viral infections and tumors.

Toxin A poison produced by or from bacteria that can cause disease.

Traveler's diarrhea A common illness affecting travelers who have three or more unformed stools in 24 hours, commonly accompanied by abdominal cramps, nausea, and bloating.

Triage The process of prioritizing sick or injured people for treatment according to the seriousness of the illness or injury.

Trigeminal ganglion A cluster of nerve cells relating to or involving the trigeminal nerve.

Trigeminal nerve Either of the fifth pair of cranial nerves that provide the jaw, face, and nasal cavity with motor and sensory functions.

Ultrasound An imaging procedure that uses high-frequency sound waves reflecting off internal body parts to create images, especially of a fetus in the womb, for medical examination.

Uremia A form of blood poisoning caused by the accumulation in the blood of urea and other waste products that are usually eliminated in the urine.

Vector An organism, such as a mosquito or tick, that transmits disease-causing microorganisms from an infected person or animal to another.

Verotoxin A poison produced by the bacterium *Escherichia coli* that inhibits protein synthesis in eukaryotic cells.

Index

abscess, 37-39
   in pelvis, 17, 18
acyclovir, 57, 58, 63
Aedes aegypti, as carrier of Dengue fever, 7
Aeromonas, 42
AIDS (acquired immunodeficiency syndrome), 13
allergic bronchopulmonary aspergillosis (ABPA), 77, 78
alveoli, 79
American Red Cross, 25, 52
amikacin, 60
amoxicillin, 3
ampicillin,
   effect on Dengue fever, 9
   effect on pseudomembranous colitis, 3
anosmia, 62
antibiotics, 1, 2, 7, 9, 12, 13, 16-18, 22, 23, 28, 36-39, 42, 43, 54, 55, 57-59, 68
   effect on pseudomembranous colitis, 3-5
antibodies, 13, 29, 30, 37, 47, 78, 79
   in response to dengue fever, 8, 9
antimony, treatment of leishmaniasis, 34
antiserum, 47
Aspergillus fumigatus, 77, 78
asthma, 75-79
azithromycin
   for asthma, 76
   for treatment of chlamydia, 17

bacillary angiomatosis, 13
Bacteroides fragilis, 4
Bartonella henselae, 12, 13
B cells, Epstein-Barr virus infection of, 29

Bell's palsy, 30
bile, 21, 22
biliary atresia, 22
bilirubin, 21
breakbone fever, 7
BUN, 71
Burkitt's lymphoma, 30

Campylobacter, 42
capreomycin, 60
cat scratch disease, 12, 13
caustics, 49
Ceftriaxone, effect on Dengue fever, 9
Cefotaxime, 54
Centers for Disease Control and Prevention (CDC), 13, 19, 45, 49, 50, 54, 60, 73, 74
cephalosporins, causing pseudomembranous colitis, 3
cerebrospinal fluid (CSF), 53, 57, 58, 63
chlamydia, 17-19, 39
Chlamydia trachomatis, 18, 39
Ciprofloxacin (Cipro), 3, 42
clindamycin, 2-4
clinical diagnosis, 53
Clostridium difficile, 3, 5
complement protein, 37
conidiospore, 78
conjugate vaccine, 55
CT scan, 43
   for meningitis, 28
   of large intestine, 1
   pelvic exam, 17
cupping, 49

Dengue fever, 9
Dengue hemorrhagic fever, 9, 10
dexamethasone, 54

Index

differential diagnosis, 22
DNA probe, 36
doxycycline, 68
   for bacillary angiomatosis, 13

EBV (Epstein-Barr virus) encephalitis, 29, 30
ectopic pregnancy, 16, 17, 39
electroencephalogram, 63
electrolytes, 41, 67, 68, 70, 71, 74
encephalitis, 28-30, 62-64
endemic, 10, 21, 24, 35
endothelial cell, 72
enterotoxin, 67
eosinophil, 79
epidemic cholera, 65, 69
Epstein-Barr virus, 29, 30
erythromycin, for bacillary angiomatosis, 13
Escherichia coli (*E. coli*), 4, 41, 67
O157:H7, 72, 73
extensively drug-resistant tuberculosis (XDR TB), 60

fallopian tubes, 18, 36-39
fibrin, 37, 38
fluoroquinolones
   effect on pseudomembranous colitis, 3
   effect on tuberculosis, 60

gastroenteritis, 41, 42
gastrointestinal infection, 2, 71
glandular fever, 30. *See also* mononucleosis
Guillain Barré syndrome, 30

hemolytic anemia, 72
hemolytic uremic syndrome (HUS), 72-74
hemorrhagic rash, 6-10
herpes encephalitis, 62-64
herpesvirus, 4
HIV (human immunodeficiency virus), 3, 19, 34
hydrocephalus, 59
hydrophobia, 48
hypochondrium, 49

immunocompromised, 77
immunoglobulin E (IgE), 77
intussusception, 44
itraconazole
   treatment of allergic brochopulmonary aspergillosis, 79
   treatment of leishmaniasis, 34
isoniazid, 60

kanamycin, 60
Keflex, 3
Koplik's spots, 22, 23

Leishamnia, 32-35
   brasiliensis, 33
   guyanensis, 33
   major, 35
   mexicana, 33, 35
   panamensis, 33
   tropica, 33, 34
leishmaniasis
   cutaneous, 34
   visceral, 33-35
leukocytes, 8
Levaquin, 3
lumbar puncture, 28, 53, 62

macrophages, 8, 32
meninges, 7, 52-54, 57, 59, 62
meningitis, 64
   and *Streptococcus pneumoniae*, 52-55
   and tuberculosis, 56-58
   related to Epstein-Barr virus, 27, 28, 30
meningococcemia, 7
meningococcus, 8
metronidazole, for treatment of pseudomembranous colitis, 4
mononucleosis, 27-30
mosquitoes, as carriers of Dengue fever, 7, 9
Mycobacterium tuberculosis, 58, 59

nasopharyngeal carcinoma, 30
Neisseria gonorrhoeae, 36, 39
neurasthenia, related to Dengue fever, 9

omentum, 37

Pan American Health Organization (PAHO), 10
pancolitis, 44
Pasteur, Louis, 49
pelvic inflammatory disease (PID), 18, 19, 37-39
penicillin
    for meningitis, 54
    pseudomembranous colitis, 3
peritoneal membrane, 38, 73. *See also* peritoneum
peritoneum, 38
platelets, 8, 29, 72
pneumococcus, 54. *See also Streptococcus pneumoniae*
pneumonia, 4, 19, 54, 62, 76, 77, 79
    related to measles infection, 23
Pneumovax®, 55
Pnu-Immune®, 55
polymerase chain reaction (PCR), 58, 59, 63
polysaccharide vaccine, 55
poultice, 49
prednisone, 79
Prevnar®, 55
prosopagnosia, 63
pseudomembranous colitis, 2, 3

rabies, 46-50
renin, 71
rifampin, 60
Rocephin, 3, 57
rotavirus, 67

Sacks, Oliver, 64
Salmonella, 42-45
septic, 1
Shigella, 42
shock, 22
    related to Dengue fever, 7-9
spinal tap, 53, 54, 57, 58. *See also* lumbar puncture

Staphylococcus aureus, 41
steroids, 29, 30, 59, 76, 78
strabismus, 51, 52
Streptococcus pneumoniae, 54

T cells, 29. *See also* T lymphocytes
temporal lobe, 62, 64
T lymphocytes,
    related to Epstein-Barr virus infection, 29
    related to measles virus infection, 23
toxin, 41, 70, 72
    cholera, 67
    pseudomembranous colitis, 3, 4
trigeminal ganglia, 62
trigeminal nerve, 62
tuberculosis, 58, 59

ultrasound examination
    of arm, 12
    of gallbladder, 57
UNICEF, 25
United Nations Foundation, 25
uremia, 72

vancomycin
    for treatment of pseudomembranous colitis, 4
    for treatment of *Streptococcus pneumoniae*, 54
verotoxin, 72, 73
Vibrio
    cholerae, 41, 66, 68
    parahaemolyticus, 42

Wearing, Clive, 64
West Nile virus, 57
World Health Organization (WHO), 9, 24, 25, 35, 49, 54, 69

Yersinia enterocolitica, 42